DESCRIPTION OF PLATE 10.

ILLUSTRATING PAPER ON THE SATELLITES OF THE MERCURY LINES.

The structure of the two yellow and the green mercury lines, photographed in the fourth order spectrum of the 40 foot spectrograph. The violet line 4358 photographed in the fifth order.

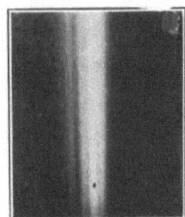 

Fig. 1. λ=5769.   Fig. 2. λ=5790.

 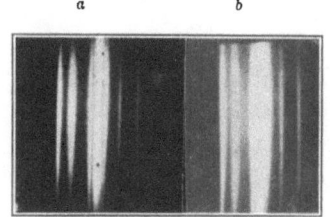

Fig. 3. λ=5461.   Fig. 4. λ=4358.

PLATE 10

COLUMBIA UNIVERSITY IN THE CITY OF NEW YORK
PUBLICATION NUMBER EIGHT
OF THE ERNEST KEMPTON ADAMS FUND FOR PHYSICAL RESEARCH
ESTABLISHED DECEMBER 17, 1904

Researches in Physical Optics

PART II

Resonance Radiation and
Resonance Spectra

By R. W. WOOD, LL.D.

Professor of Experimental Physics in the Johns Hopkins University
Adams Research Fellow of Columbia University, 1911-1914
Foreign Member of the Royal Society

NEW YORK
COLUMBIA UNIVERSITY PRESS
1919

Copyright 1919 by Columbia University Press

DOUGLAS C. McMURTRIE
THE ARBOR PRESS, INC.
NEW YORK CITY

ON the seventeenth day of December, nineteen hundred and four, Edward Dean Adams, of New York, established in Columbia University "The Ernest Kempton Adams Fund for Physical Research" as a memorial to his son, Ernest Kempton Adams, who received the degrees of Electrical Engineering in 1897 and Master of Arts in 1898, and who devoted his life to scientific research. The income of this fund is, by the terms of the deed of gift, to be devoted to the maintenance of a research fellowship and to the publication and distribution of the results of scientific research on the part of the fellow.

(For list of Adams Fund Publications see back cover page)

FOREWORD

THE researches described in the following pages were made by Professor Wood, in several cases together with collaborators, while he was Ernest Kempton Adams Fellow of Columbia University in 1912–1914, and subsequently. In this work, Part II of these Researches in Physical Optics, the important field of investigation opened by the author has been developed in the direction of more detailed study of resonance radiation and resonance spectra.

Publication has been delayed from time to time, in order to include new material and, more recently, on account of conditions arising from the war. In the absence in Europe of the author as the volume goes to press, the editors of the series venture to express for the author his obligation and thanks to his several collaborators and to all who have assisted in his work. To the *Astrophysical Journal* thanks are due for cuts used in illustrating the fifth paper.

NEW YORK, *April, 1919*

TABLE OF CONTENTS

		Page
1.	Plane Grating Spectrographs of Long Focus	1
2.	The Resonance Spectra of Iodine	6
3.	Resonance Spectra of Iodine	20
4.	The Series of Resonance Spectra With M. Kimura	38
5.	Band and Line Spectra of Iodine With M. Kimura	51
6.	Zeeman—Effect for Complex Lines of Iodine With M. Kimura	68
7.	A Photometric Study of the Fluorescence of Iodine Vapor With W. P. Speas	77
8.	The Magneto-Optics of Iodine Vapors With G. Ribaud	85
9.	The Fluorescence of Gases Excited by Ultra-Schumann Waves With G. A. Hemsalech	95
10.	A Further Study of the Fluorescence Produced by Ultra-Schumann Rays With C. F. Meyer	106
11.	Scattering and Regular Reflection of Light by an Absorbing Gas	117
12.	Separation of Close Spectrum Lines for Monochromatic Illumination	147

		Page
13.	Photometric Investigation of the Superficial Resonance of Sodium Vapor . . . . *With L. Dunoyer*	155
14.	The Separate Excitation of the Centers of Emission of the D Lines of Sodium . . . . *With L. Dunoyer*	166
15.	Resonance Radiation of Sodium Vapor Excited by One of the D Lines . . . . *With F. L. Mohler*	174

No. 1

Plane Grating Spectrographs of Long Focus

In the Researches in Physical Optics, Part I, page 26, I gave a brief description of a plane grating spectrograph of the Littrow type with a focal length of forty feet. The instrument was installed at my summer laboratory at East Hampton and was used in the study of the remarkable absorption spectrum of iodine vapor, which, as I pointed out at the time, is by far the best test for the resolving power of large gratings which we have.

This instrument gave a good deal of trouble as a result of the circulation of air currents in the tube, and performed at its best only late at night when a temperature equilibrium had been established.

To get rid of this difficulty I determined to put the entire apparatus underground, and though this expedient introduces one or two new difficulties, they are of minor import, and I have very little fault to find with the present arrangement.

The tube of the instrument was made of six-inch earthenware drain pipes laid in a trench by the local mason. To insure absolute straightness I placed a small heliostat in the trench, and marked with a beam of sunlight the axis of the tube, each section of pipe being adjusted by the central ray. When completed, the inner surface was as smooth and straight as the bore of a rifle, much to the surprise of the mason, who was misled by the somewhat irregular appearance of the outer surface which resulted from the variable thickness of the walls of the sections. The tube enters a small cement lined cellar, which I built for the purpose, about four feet below the surface of the ground. The slit tube and plate holder are mounted on a six-inch tube of galvanized iron which slides into the last section of the drain pipe, the focussing being accomplished by pushing in or drawing out the tube. This rather crude device was found to be perfectly satisfactory, though a rack and pinion movement would of course be better. The other end of the tube opens into a cubical chamber of cement, which was cast around the flared end of the last section of the tube, the light wooden box which served as a form for the inner surface of the chamber being knocked to pieces after the cement had set. No form, other than the hole in the ground, was used for the outer surface. After the cement had thoroughly dried out the inner surface was painted with a thick coating of a water-proof asphalt compound, and a dish of calcium chloride used to keep the air dry.

This cement chamber contained the lens, grating and electric motor for rotating the grating support. It was closed by a wooden cover and buried under two feet of soil, after all adjustments had been made. I expected that this would prevent the circulation of air in the tube, and was somewhat surprised to find a steady draught of cool air issuing from the mouth of the tube in the cellar, and the definition very poor. It is obvious that if warm air enters the tube and flows down it the constant temperature to which we have brought the tube avails us nothing. This circulation I attribute to a slight difference in the barometric pressure between the outside and the inside of the building resulting from the wind. It was present in the earlier instrument, except on very calm nights, but I was surprised to find that it made itself felt through the thick layer of soil which covered the grating box.

It was obvious that the cement chamber must be hermetically sealed, and it was desirable to have the interior of easy access. This was accomplished by providing it with a cover of galvanized iron fastened permanently to the walls with Portland cement and perforated with a hole eighteen inches in diameter. A shallow trough of galvanized iron was soldered around the rim of the hole, into which the rim of the circular cover (made like the cover of a tin pail fitted. A pint of heavy lubricating oil, poured into the trough, completed the seal.

The definition, which was very bad, before the seal had been completed, became perfect as soon as the oil was poured into the trough.

The lens which I used during the summer of 1913 was the same as that used in the arrangement of the previous year, a forty-foot coronagraph lens kindly loaned by Professor Campbell of the Lick Observatory. The only trouble which I experienced was the occasional condensation of moisture between the lenses.

The images of the illuminated slit, reflected from the four surfaces of this lens were thrown out of the field by tilting the lens, which is, I believe the approved method. This reflected light does very little harm when photographing discontinuous spectra, but when working with the solar spectrum, especially in the higher orders, it may well very happen that the plate receives much more light from the surfaces of the lens than from the grating. On discussion the matter with Mr. Twyman, of the Hilger Company, he suggested that they make me a cemented lens figured so as to give two real images of the slit by reflection, these images to be covered with narrow vertical wires. In this way the inclination of the lens could be avoided, and the definition perhaps improved. By employing a cemented lens two reflecting surfaces are abolished, and the trouble with the moisture is done away with. I accordingly had a lens made of this type. Its focal length is forty-two feet, and the two images formed by reflection are at distances of six feet and eighteen feet from the lens. This lens was substituted for the one borrowed from the Lick Observatory, and adjusted so as to show the reflected images of the slit at the center of the tube. The problem now presented itself of getting the vertical wires into the tube at exactly the right place. The brighter of the two images (as seen from the slit) is of course the long focus one which is nearly at the center of the tube. The best plan appeared to be to mount the wire on a flat plate and draw it down the tube by a wire. This necessitated getting a string through the instrument, and the family cat, which was utilized on former occasions for removing spider webs from the long tube, having died without issue, I was obliged to spend some time in learning how to throw an iron ball attached to a long thread through the drain pipe, without having it cut the thread. This was finally accomplished, and a copper wire was then drawn through by a thread. Once having a wire through the tube, spider webs can be removed at any time regardless of the cat's disposition, and the screening wire brought into the desired position.

The eclipsing wire was about half a millimeter in diameter, mounted vertically on a lead plate and was brought into coincidence with the image at the center of the tube, the fine setting being made by the adjusting screws of the lens cell. There remained now the other image formed by the reflecting surface of shorter focus. Though it was not very bright it seemed best to remove it and another wire was introduced by digging down to the pipe, drilling through the wall, and cementing a short piece of brass tubing in the hole. The wire was mounted on a brass rod which rotated in the tube, and was bent through two right angles in such a way that by rotating the rod the vertical portion of the wire crossed the field. To my great surprise I found that as the wire was brought towards the center of the field, its edges became strongly illuminated and when it exactly covered the image of the slit, it actually sent to the plate three or four times as much light as the image which it screened. Long experience with the "Schlieren-methode" made me recognize the trouble as analogous to the diffraction effects shown by the edges of opaque screens when examined by the Schlieren apparatus, which is essentially the Foucault arrangement for testing lenses or mirrors. The rays are obliged to pass by the edge of the wire twice, and the diffracted energy is not brought accurately to a focus at the long focus image and consequently gets by the screening wire.

This difficulty could hardly have been predicted, or rather one would scarcely have expected to find that the wire, when in position, would send more light to the plate than it cut off. As there appeared to be no way of remedying the trouble, the wire at the center of the tube was removed, and the long focus image thrown out of the field by tilting the objective very slightly in the vertical plane. This did not impair the definition, as the tilt was extremely small on account of the great distance of the image from the lens. The short focus image was covered by the adjustable wire as described above.

The cement chamber containing the grating was covered with the small wooden house described in the earlier paper. This is provided with double walls, and the shingles on the outside are nailed to cleats, so that there is an air space between them and the boards of the house.

The present disposition of the apparatus has given excellent satisfaction, though I had a little trouble late in the summer when the first cold nights came, from condensation of moisture on the surface of the lens which was in contact with the air of the tube. This occurs only when the temperature of the cement chamber falls below that of the ground at the depth at which the pipe is buried. I found the definition perfect at night with exposures of two hours for some of the mercury lines, but for very long exposures say of twenty-four hours, it would be necessary to control the temperature of the grating chamber with a thermostat. If a suitable building is available there is no necessity of burying an instrument of this type underground. The method was adopted in the present instance merely because it appeared to be the only solution of the problem for a laboratory housed in an old barn. Three things must be prevented if the full resolving power of a large grating is to be obtained: (1) mechanical vibration; (2) temperature changes of the grating in excess of one or two tenths of a degree centigrade; (3) circulation of air in the tube.

In regard to the latter condition it is probable that no trouble from this source will be found if the entire instrument is mounted in a single room. If there is the slightest temperature variation along the walls of the tube, a very feeble air current gives rise to striæ which ruin the definition.

In the case of instruments installed in the usual manner in laboratories the use of a toluene or mercury thermostat capable of holding the temperature of the grating constant to within $0.2°$ C. is imperative for exposures of over an hour.

If an instrument of this type is to be installed in a laboratory it is my opinion that the best plan would be to support the grating mount and the slit and plate-holder on piers preferably built up from the ground in the basement or still better in a sub-cellar. One instrument which I recently assembled on the third floor of our new laboratory has the piers built on the cement floor. It was found however that the image of the slit drifted gradually first to the right and then to the left in a very irregular manner, a change of position being noticeable in half an hour on some occasions. The trouble disappears late at night. It is due to the warping of the building as the sun moves from one side to the other, and to temperature changes in the interior. If it is imperative to install the instrument in a room above the basement, a long I beam must be used as a base, the ends being carried on piers.

No. 2

The Resonance Spectra of Iodine

In *Researches in Physical Optics*, Part I, two papers were published on the remarkable spectra emitted by iodine vapor at room temperature in a highly exhausted tube, when excited to luminosity by the green and yellow rays of the mercury vapor lamp.

During the three years which have elapsed since the appearance of this publication the work has been continued with improved apparatus and methods, and more powerful spectrographs, until at last it has been found possible to photograph some of the brighter groups of lines in the fourth order spectrum of a seven-inch plane grating with a lens of three meters focus, *i. e.*, with a resolving power just sufficient to separate the seven absorption lines of iodine which are covered by the broadened green line of the quartz mercury arc.

To make the present paper a complete presentation of the subject it will be necessary to incorporate in it some of the material given in the earlier paper, as the continued study of the subject has made it necessary to modify some of the views originally held.

GENERAL NATURE OF RESONANCE SPECTRA

The discovery of resonance spectra which I announced in 1904, and their study, which I have outlined in a series of papers which have appeared in the Philosophical Magazine since that date and the present time, have furnished a new method of attacking the problem of the nature of molecular radiation. Resonance spectra arise in the following manner. The vapors of sodium, potassium, iodine and certain other elements and compounds, such as the oxides of nitrogen, exhibit absorption spectra consisting of thousands of very fine lines grouped together with more or less regularity into bands, which in certain cases appear to be arranged according to a certain law as has been observed by Deslandres and others. Some of these vapors when illuminated with a powerful beam of white light become luminous (*i. e.*, fluoresce) and if we examine the emitted light with a spectroscope we find that the spectrum, roughly speaking, is the complement of the absorption spectrum, that is, the dark lines of the one are replaced by bright lines in the other. The simplest way of explaining this phenomenon would be to assume that each absorption line removes from the incident beam the frequency corresponding to its own, and reëmits this frequency laterally in all directions. This explanation was found to be insufficient, for experiment showed that if sodium vapor was illuminated with light comprised between rather narrow limits of wave-length, cut out from a continuous spectrum, the fluorescent spectrum covered practically the same spectrum range as when white light was employed. The experiment was then tried of illuminating the vapor with monochromatic light obtained from a metallic arc, allowing only light corresponding to one of the spectrum lines to reach the vapor. *The spectroscope now showed that the vapor was emitting a series of isolated bright lines spaced at nearly equal distances along a normal spectrum.* One line coincided in position with that of the exciting line, from one to three appeared on the short wave-length side of it, and the remainder, to the number of ten or a dozen, on the long wave-length side. To spectra excited in this way I gave the name resonance spectra, since they appeared to be originated by a resonance of the absorbing molecule for a definite frequency, the energy abstracted from the incident beam being for the most part distributed among other frequencies by a mechanism within the molecule, the nature of which has not been definitely determined up to the present time, though two or three promising hypotheses have been offered, as I shall show presently. The study of these resonance spectra was attended with great experimental difficulties, since the intensity of the light emitted under monochromatic stimulation was so faint that exposures varying from eight to twenty hours were necessary. For this length of time it was necessary to keep the metallic arc, which requires constant attention, burning steadily. Moreover the vapor of the metal had to be kept away from the glass walls of the vessel which contained it, involving further difficulties, which it is not necessary to go into, but which will be found described in my earlier papers. In the autumn of 1910 I discovered the resonance spectrum of iodine vapor excited by the monochromatic radiations from the mercury arc burning in a quartz tube, and at once the study of these spectra became very simple since these lamps can be run continuously several thousand hours without attention, and the iodine vapor can be enclosed in glass bulbs or tubes, which require no heating, as the most favorable density of the vapor turned out to be that which obtains at room temperature.

THE ABSORPTION SPECTRUM

The absorption spectrum of iodine has been studied with the forty-two foot plane grating spectrograph described in the previous paper. It is made up of a large number of fluted bands, and resembles in its general appearance the channeled absorption of sodium vapor, for in both cases the bands at the long-wave length end are quite regular in their appearance, while at the short wave-length end they become more or less confused. As I have already said the number of absorption lines, which collectively form the banded spectrum of iodine, has been greatly underestimated. Sunlight from the heliostat was passed through a large exhausted bulb containing a few small crystals of iodine and focused upon the slit of the instrument. The absorption spectrum seen under these conditions presented a most wonderful appearance, nearly the entire visible spectrum being filled with thousands of lines. As I have said, I found seven sharp and beautifully resolved lines within the region covered by the green emission line of mercury. The total width of the line was 0.4 A.E. and we have at this rate eighteen lines to the Ångström unit or about 36,000 lines in all. There were, however, groups containing lines much closer together than the seven lines just enumerated, which were still unresolved by the grating and numerous broad dark bands undoubtedly made up of unresolved lines. Over one hundred lines have been counted between the D lines of sodium. This circumstance, together with the fact that the lines are much closer together in the red, orange and yellow region, makes me feel certain that there are upwards of 40,000 lines in this remarkable absorption spectrum.

The wave-lengths of the seven lines which were observed within the green mercury line (furnished by the quartz mercury arc) were very carefully measured with the eye-piece micrometer, with reference to the wave-lengths of the components of the mercury line seen with a low temperature arc.

They were subsequently measured from photographs taken with the 40-foot spectrograph, with reference to the main component of the green mercury line, which falls midway between two of the iodine absorption lines, and the wave-length of which is 5,460.7424.

The wave-lengths of the seven iodine absorption lines are as follows:

$$5,460.966$$
$$.910$$
$$.973$$
$$.873$$
$$.768$$
$$.716$$
$$.640$$
$$.579$$

A portion of the absorption spectrum in the region of the green mercury line is reproduced on Plate I, Figure a, the seven lines enumerated above being enclosed between the two arrows, and when it is remembered that the entire portion of the spectrum reproduced embraces a range of the spectrum not much greater than the distance which separates the D lines of sodium, the frightful complexity of these absorption spectra becomes evident.

I have examined the absorption spectrum of sodium with this very powerful apparatus and find that it is equally complex, the distance between the lines being about the same as in the case of iodine. Sodium however exhibits only a single faint line within the green mercury line, and as this appears only when the vapor has a considerable density, the mercury arc is incapable of stimulating this vapor to appreciable fluorescence.

The bromine absorption spectrum is very similar in appearance to that of iodine. Further reference will be made to this when I come to the subject of the use of bromine vapor as a ray filter, for modifying the intensity distribution in the green mercury line.

THE RESONANCE SPECTRUM

The form of tube finally adopted for the study of the iodine fluorescence is much more efficient than the large bulbs used in the earlier work. The tube is prepared in the following way: A piece of thin walled tubing about 3 cms. in diameter and 30 cms. long is blown out at one end in the form of a bulb 4 cms. in diameter. It is important to avoid having a thick drop of glass form in drawing down the tube previous to blowing, as this forms a lens on the surface of the bulb. The tube should be carefully cleaned and dried and furnished with a short lateral branch, of quarter inch tubing.

This is used for controlling the density of the iodine vapor, as it is often advantageous to work at densities less than that which obtains at room temperature. The other end of the large tube is now drawn down and a smaller tube fused on for the exhaustion. A few small flakes of iodine are introduced, and the tube drawn down and bent into a U which is immersed in liquid air during the exhaustion, to prevent the entrance of iodine into the pump. If liquid air is not available, solid CO_2 answers as well. It is important to have a high vacuum in the tube, and if the iodine crystals are at the bottom of the lateral branch, the large tube can be heated with a Bunsen burner. It is a good plan to keep the end of the lateral branch at a low temperature to prevent evaporation of the iodine, and continue the pumping for fifteen or twenty minutes, after which the tube is sealed off from the pump.

For the excitation of the resonance spectra either the Cooper-Hewitt mercury lamp or the quartz mercury arc may be used. The latter is the better for demonstration purposes, as the fluorescence is much brighter. An image of the lamp is focussed along the axis of the iodine tube as near the bulb as possible, by means of a large condensing lens of short focus: the fluorescence is observed 'end-on' through the bulb.

For excitation by the Cooper-Hewitt lamp it is necessary only to mount the iodine tube parallel to, and almost in contact with, the tube of the lamp. In the present work I have used a very large and powerful Cooper-Hewitt glass lamp, the tube having an internal diameter of 5 cms., and carrying a current which can be varied from three to fifteen amperes. A novel method of illumination was used in the case of this lamp, which appears to be the most efficient found up to the present time. It is illustrated by Fig. 1. Three enormous cylindrical reflectors were made from the curved glass plates used for the corners of shop windows. These were furnished silvered by a mirror-glass company at a very small cost, and when mounted as shown formed a nearly complete cylindrical reflecting shell over half a meter in diameter and in length. The exhausted tube containing the iodine and the large Cooper-Hewitt lamp were mounted side by side along the axis of the reflecting cylinder, the whole arrangement forming a sort of 'light furnace,' in which the iodine vapor glowed with great brilliancy.

One disadvantage of this method of illumination is that the tube is considerably heated by the lamp and the resonance spectrum is modified by the absorption of the dense iodine vapor. This trouble could be remedied by keeping the end of the lateral branch outside of the 'furnace', for the density of the vapor is

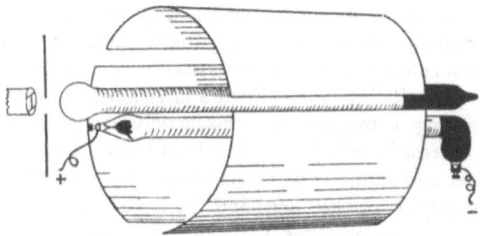

FIGURE 1

determined by the temperature of the coldest part of the tube. A simple form of light furnace will be described later on.

Further control of the method of excitation was obtained by passing the light of the mercury arc through a large glass bulb 40 cms. in diameter, exhausted save for enough bromine vapor to give the transmitted light a good yellow color. The bromine vapor has certain lines in coincidence with iodine absorption lines, and by its use as a ray-filter we may remove certain frequencies from the green mercury line and in consequence throw certain iodine lines out of action.

In my previous communication I showed that the green line of the mercury-vapor lamp had a width sufficient to enable it to cover from two to seven of the fine absorption lines of iodine. In the case of the Cooper-Hewitt glass lamp the green Hg. line was so narrow that only one absorption line was excited, while in the case of the quartz lamp, running at a high temperature, the line was so broadened that seven absorption lines were covered. In the latter case the resonance spectrum consisted of groups of lines occurring at regular intervals along the spectrum, and to this type of stimulation I gave the name 'multiplex excitation'. I predicted at this time that the appearances of the groups of lines could be profoundly modified by running the quartz lamp under different conditions as to temperature and voltage drop across the terminals, as in this way the width of the exciting line could be varied, and the number of absorption lines excited easily controlled.

In the more recent work the iodine vapor was excited by the light of a quartz mercury-vapor lamp operating with various resistances so as to show a potential drop across the terminals varying from 25 volts to 165 volts. The tendency of the lamp to go out when running with a large resistance in circuit with it, was overcome by including in the circuit a large amount of self-induction, consisting of two coils of very heavily insulated wire with laminated cores, for which I am indebted to Mr. Cooper-Hewitt. A voltmeter and ammeter was used in each case to determine the watt consumption of the lamp, and the appearance of the green exciting line in coincidence with the absorption spectrum of iodine was observed during each exposure, in the fourth order spectrum of a 6-inch grating with a lens of 3 meters focus. In addition to this, a valuable series of photographs of the green line in coincidence with the absorption spectrum was made for me by Dr. J. A. Anderson, with the large spectrograph of the Mt. Wilson Solar Observatory. This instrument has a focal length of 75 feet and is furnished with a Michelson grating. It appears to have about the same resolving power as my East Hampton spectrograph, though its greater focal length enables it to yield photographs of much more satisfactory appearance, the definition in the case of my instrument being marred somewhat by the grain of the plate.

The method of controlling the excitation which I adopted necessitates the use of two powerful spectrographs, one for photographing the resonance spectrum, the other for studying the appearance of the green exciting line during the exposure and determining which of the iodine absorption lines are excited by it.

The adoption of this plan has thrown much new light upon the genesis of resonance spectra, and further improvements in the methods of utilizing the exciting light have made it possible to photograph the resonance spectra in the fourth order spectrum of the large plane grating which I use in the 42-foot spectrograph, though with a lens of shorter focus (3 meters) and with an exposure of only 24 hours. When it is remembered that in the first study of the iodine fluorescence an exposure of 24 hours was necessary to secure a satisfactory photograph with a Hilger one-prism spectrograph, the immense gain in efficiency is apparent. I can now secure good photographs of the resonance spectrum in the first order spectrum of the grating in 30 minutes.

The work of the past year has furnished some very definite problems in molecular mechanics for the theoretical physicists to solve, and I feel that, for the first time, I am now able to furnish some very exact knowledge of the nature of these remarkable spectra, which it is most important to account for by theory.

Owing to the complexity of the resonance spectra which arise from what I have called multiplex excitation, when the exciting arc line is sufficiently broadened to cover a number of the iodine absorption lines, it will be best to consider first the simpler spectra which arise when the lines of the Cooper-Hewitt mercury lamp are used for the stimulation.

The green line in this case is so narrow that it covers but one of the seven or more absorption lines which are covered by the green line emitted by the quartz mercury arc operating at a high temperature. For the present we will neglect the feebler stimulation of the vapor by the light of the satellites of the green line. Referring to Plate I, Figure a, we find between the two arrows the seven iodine absorption lines above referred to. These lines we will number '1 to 7' beginning from the left. It is the stimulation of line No. 3 with which we are now concerned. The scale of wave-lengths for the absorption spectrum is immediately below it. The green mercury line ($\lambda = 5460.74$) lies on the right hand edge of absorption line 3 and is nearly equal in width to the distance between lines 3 and 4. The resonance spectrum resulting from the excitation of this absorption line consists of a series of doublets, of which twenty-three members have been photographed, the last lying in the extreme red.[1] Using the previous nomenclature we shall speak of the doublet which arises at the point of excitation as the group of 0 order, and the succeeding doublets of increasing λs as groups of 1st, 2nd, 3rd, etc., order.

Groups to the left (short wave-length side) of the excited line will be designated as -1, -2, etc. On Plate II, Figure j, will be found a reproduction of the groups from 0 to 5 inclusive, made in the first order spectrum of the large spectrograph (7-inch plane grating and lens of 3 meters focus). The comparison spectrum is that of the molybdenum arc.

The doublet of 0 order is made up of a line resulting from the reëmission, without change of wave-length of the light absorbed by absorption line 3 of iodine. This is the very black line at the extreme left of the spectrogram. Close to this line, on the right is the companion line, the two forming the doublet of 0 order. On Plate I, Figure a, this companion line ($\lambda = 5462.22$) would come in the position indicated above the spectrogram, the doublet of 0 order having been drawn in its proper position. Whether or not it is in exact coincidence with an absorption line can be told with certainty only by photographing the doublet in coincidence with the absorption spectrum with a spectrograph of resolving power sufficient to clearly separate the iodine lines. This will probably be possible, as the resonance spectrum excited by the quartz arc has already been photographed in the fourth-order spectrum with a lens of three meters focus. With a lens of double this focus clear separation of the absorption lines would be obtained. The circumstance that the members of the doublets are absorbed by iodine vapor in very different degrees, makes it seem probable that the resonance lines are not all in coincidence with absorption lines.

For example if we reduce the density of the iodine vapor by cooling a portion of the tube to zero, the two members of the first order doublet are of equal intensity. If we warm the tube to 30° or so, or pass the light from the cooled tube through a large globe containing iodine vapor, the right hand member of the doublet (longer λ) disappears almost completely.

In the upper photograph on Plate II its intensity is only about one-third that of the other member. There is little or no

[1] This number has recently been raised to twenty-seven. See second part of paper.

difference in the absorbing power of the vapor for the two members of the doublet of the third order, but for the fourth order the member of *shorter* wave-length is very much weakened by iodine absorption. This doublet is accompanied by another doublet lying to the left on the Plate, which is the doublet of —1 order due to the excitation by the yellow line 5790.6.

The doublets themselves are of variable intensity and some are missing entirely. In the case of the second order group a very faint doublet appears, but it does not appear to belong to the series as it is shifted about 0.8 ÅU. (towards short λs) from the position which it should occupy.

The wave-lengths of the doublets have now been determined to within 0.1 ÅU. certainly, and probably to within .05 ÅU. and the law which governs their spacing along the spectrum determined. If the frequencies are taken for, say, the left-hand member of each doublet, we have the lines represented by the following formula.

(1) $\frac{1}{\lambda} = 183075 - 2130m \times \frac{(m-1)m}{2} 12.2 \begin{pmatrix}\text{For excitation by}\\ \text{Hg } 5460.74\end{pmatrix}$

Reducing the wave-length 5460.74 to vacuum (5462.23) we have in the formula $\frac{1}{\lambda} = E$ reciprocal of (vacuum) wave-length of left-hand member of doublet of order m.

$183075 = \frac{1}{5462.23}$

2130 = difference between $\frac{1}{\lambda}$ for 0 order and $\frac{1}{\lambda}$ for first order

12.2 = constant second differences of $\frac{1}{\lambda}$

If we represent frequencies by $\frac{1}{\lambda}$ and call $x = 2130$ and $n = 12.2$ we have the frequencies of the left-hand members of the doublet series represented by (if $a = 183075$, the frequency of the green mercury line).

(2) $a, a-x, a-2x+n, a-3x+3n, a-4x+6n \ldots$ etc. The coefficients of n in any term is the sum of the coefficients of x and n in the preceding term. The law holds for frequencies

and not for wave-lengths. If it held for wave-lengths we should have the lines as follows:

$$a, \ a+x, \ a+2x+n, \ a+3x+3n, \text{ etc.}$$

in which $a = 5460.74$; $x = 5525.04 - 5460.74$ and $n =$ second difference of λ s or expressing it in words, the distance between any line and the line *above* it would be equal to the distance of the line *below* it, plus a constant small increment, 'n'. The wave-lengths of the doublets are given in the following table. They are from measurements made on plates taken with the three meter spectrograph. Comparison spectrum molybdenum, and International scale used. The values were reduced to vacuum, and these corrected λs were used in applying the formula.[2]

Doublet Series Excited by $\lambda 5460.74$ Hg. Cooper-Hewitt Lamp.

m.	λ obs.	Width of doublet	λ (cor to vacuum)	$\frac{1}{\lambda^1}$ (obs.)	$\frac{1}{\lambda^1}$ (cal.)	Difference
0	5460.74 5462.22	1.48	5462.23 63.71	183075	183075	0
1	5525.04 5526.58	1.54	5526.55 28.09	180944	180945	+1
2	5589.58 5591.70	Does not belong to series	5591.10 93.23	178856	178827	0
3	5657.17 5658.81	1.64	5658.71 60.35	176719	176720	−1
4	5725.01 5726.64	1.63	5726.57 28.20	174625	174626	−1
5	5794.07 5795.79	1.72	5795.65 97.37	172544	172544	0
6	5864.45 5866.17	1.72	5866.05 67.77	170473	170474	−1
7	5935.96 37.87	1.91	5937.58 39.49	168419	168417	+2
8	6008.93 6010.76	1.83	6010.57 12.40	166374	166372	+2
9	Missing					
10	6159.14 6160.90	1.76	6160.82 62.58	162316	162318	−2

[2] A more accurate table and test of the formula will be found further on in the paper.

Calculations by formula $\frac{1}{\lambda} = 183075 - 2130.6m + 12.2 \frac{(m-1)m}{2}$

The doublets which have been under discussion I feel very sure arise from the excitation of absorption line No. 3 in the group of seven which are covered by the broadened green line emitted by the quartz arc running at a high temperature. On Plate I, Figure a, we have a photograph of the green line of the Cooper-Hewitt lamp in coincidence with the absorption spectrum of iodine. The main line lies between absorption lines 3 and 4, rather nearer the former. This makes it appear probable that the doublets arise from the excitation of this line.

It is of course possible that the simultaneous excitation of the two absorption lines 3 and 4 is responsible for the doublets, but I do not think that this is probable. If now we excite the vapor by the light of the quartz mercury arc, running with a potential difference of 60 volts between its electrodes the resonance spectrum appears as in Plate II, Figure k. Two new lines appear to the left of the doublets, in the groups of order 0, 1, and 3. The left hand one (*i. e.* the one further removed from the doublet) appears to be due to the stimulation of absorption line No. 4. We have in this case I believe a series of doublets as before, the right-hand member of which coincides (nearly) with the left-hand member of the doublets excited by line 3, the *companion line lying to the left however*. We can discriminate between what I call the companion line and the main line of each doublet in this way: the main lines form a series expressed by the formula previously given, of which one member (0 order) coincides with the absorption line which is excited.

If the fluorescence is excited by the quartz arc at 115 volts we find complicated groups of lines instead of the simple doublets excited by the Cooper-Hewitt lamp. These groups are very similar in their arrangement of lines and the circumstance that we find a similar group of 0 order furnishes us with a clue as to how they originate. They are shown on Plate II, Figures l and m. The orders 0, 1, 3 are almost identical in appearance, and the orders 5, 6, and 8 are sufficiently like them to enable us to identify some at least of the corresponding lines. (Figure o.)

These groups originate in the following way: The seven absorption lines which are covered by the broadened green mercury line are simultaneously excited, and the vapor emits these seven wavelengths without change. These lines we may call the R. R. lines (resonance radiation). Each one of these is moreover the first member of a series such as is expressed by the formula previously given. The R. R. lines are not resolved by the spectrograph employed in photographing the resonance spectra and consequently appear superposed. But each one is accompanied by one or more companion lines, lying to the right or left, and it is these companion lines which form the group of 0 order. The actual width of the group of seven R. R. lines is only about one-thirtieth of the width of the group formed by the companion lines.

THE BAND SPECTRUM AND THE LINE SPECTRUM: EFFECT OF REDUCING THE DENSITY OF THE IODINE VAPOR

In the course of the experiments made to determine the effect of the absorption upon the resonance spectrum some interesting observations were made. As I have shown in previous papers, the band spectrum is developed by the admixture of helium or gases of its group, at a pressure of a few millimeters, with the iodine vapor. The intensity of the resonance groups diminishes gradually and that of the band spectrum increases proportionally as the pressure of the helium increases. The same thing is true, though to a less extent, with other gases such as nitrogen, the total intensity of the light being much less, however, for reasons given in the papers published by Wood and Franck. To reduce the absorption element as much as possible, the rear end of the iodine tube was packed in ice, or in a mixture of ice and salt, which reduced the density of the iodine vapor to a small fraction of its value at room temperature. It was found that in this case the band spectrum was quite pronounced. With the iodine vapor in a high vacuum, one obtains always a faint trace of the band spectrum if a very long exposure is given, but it was much stronger in the case of the cooled tube. The hypothesis was made that this resulted from the circumstance that the band spectrum was more strongly absorbed by the iodine vapor than the lines of the resonance groups, some of which as we have seen are absorbed scarcely at all. This was tested by passing the light from the frozen tube through a bulb containing iodine vapor at room temperature, before its entrance into the spectroscope. The band spectrum at once disappeared, showing that the hypothesis was in all probability correct. The resonance groups were uninfluenced by the cooling of the tube, except that they became fainter.

EXAMINATION OF THE VAPOR FOR PHOSPHORESCENCE

An attempt to detect a possible finite duration of the light emitted by the vapor after shutting off the exciting beam, was made by focussing an image of the sun at the center of a swiftly moving stream of iodine vapor. Two glass bulbs were joined by a tube which projected several centimeters into one of the bulbs. The iodine crystals were introduced into the other bulb and the whole system highly exhausted and sealed from the pump. On cooling the first bulb by the application of a pad of cotton-wool wet with liquid air, the iodine vapor in this bulb immediately condensed, forming a very high vacuum into which rushed the vapor, continuously formed from the crystals in the other bulb. The solar image was formed just at the mouth of the tube, but no prolongation of the fluorescent spot could be detected, as would be the case if the moving jet of vapor remained luminous after passing through the focus. A paper by Mr. F. S. Phillips has appeared in the Proc. Roy. Soc. (ser. A, vol. lxxxix.) since the completion of my work describing similar experiments with mercury vapor, which showed very persistent luminosity. It is probable that the fluorescence of mercury vapor results, in part at least, from the return to the atoms of electrons expelled by the action of the light-waves, for there is no trace of polarization of the light. The fluorescence of iodine, sodium, and potassium vapors is strongly polarized, however, as I have shown in previous papers, and the polarization is for the light of the complete resonance spectrum (*i. e.* not confined to the R. R. line). This makes it seem probable that the fluorescence results directly from disturbances set up in the atom, and not from the falling back of electrons. On this hypothesis we should expect phosphorescence to be shown only by mercury vapor, for it is inconceivable that vibrations set up in the electron system of an atom could persist long enough to be detected. If we have, however, something analogous to dissociation and recombination, it is clear that phosphorescence may be apparent if only the latter process is sufficiently delayed.

No. 3
Resonance Spectra of Iodine

As has been shown in the previous communication, the vapor of iodine in vacuo, when excited to luminosity by the light of the Cooper-Hewitt mercury lamp (glass) emits a spectrum consisting of a series of doublets, with a separation of about 1.5 ÅU. very regularly spaced along the spectrum and separated by intervals of about 70 ÅU. These intervals increase gradually, however, as we pass away from the green mercury line, at which point the doublet series has its origin, until, in the extreme red, the distance between the last two doublets observed is about 102 ÅU., and the separation of the components of the doublet has increased to 2.8 ÅU. By the use of dicyanine plates the series has been followed to its termination at wave-length 7685 and the wave-lengths of the seven new doublets accurately measured. The doublets are not all of uniform intensity, and some are missing entirely, and it is the connection between this circumstance and the way in which the doublet series is related to the band absorption spectrum, that is the most interesting point brought out by the recent investigations. By varying the conditions of the experiment it has been found possible to excite by the green mercury line not only the doublet series, but a simplified system of fluted bands, few in number and regularly spaced if the iodine is in vacuo, increasing in number and complexity if a gas of the helium group is mixed with the iodine, or if more than a single iodine absorption line is excited by the mercury lamp. It is probable that the lines forming the doublets are themselves constituents of the fluted bands, and the transfer of energy from one part of the vibrating system to another, as a result of collisions between iodine and helium molecules, enables us to build up, so to speak, the complicated system of fluted bands shown in the absorption spectrum, from a number of simpler systems which can be excited separately. This constitutes a very great advance in the analysis of band spectra, and brings us a step nearer to the point at which we can picture some idea of the vibrating mechanism.

In the more recent work a method of illumination has been employed which is distinctly superior to any previously used, and as it is well adapted to purposes of demonstration I shall describe it in some detail. The iodine tubes which I now employ are of soft glass, about 40 cms. long and 3 cms. in diameter. One end is blown out into a thin bulb, taking care to avoid having the thick drop near the center of the bulb. This is best accomplished by drawing off the tube in an oblique direction, which brings the drop—formed by the melting down of the pointed end—well to one side. If this is not done the drop is apt to form a small lens on the surface of the bulb exactly on the axis of the tube.

The other end is drawn down, and a few flakes of iodine introduced into the tube. It is a good plan to provide the tube with a lateral branch, by which the density of the vapor can be controlled, though this is not necessary for demonstration purposes. The iodine flakes are now brought into the bulb, or to the bottom of the lateral tube, and the tube joined to a Gaede pump, interposing a U tube immersed in liquid air or solid CO_2, or a tube filled with caustic potash, to keep the iodine out of the pump. During the exhaustion it is a good plan to heat the walls with a bunsen flame, except where the iodine is located. Then allow the tube to cool down to the temperature of the room, and heat the portion where the iodine is located. The flakes will sublime rapidly and crystalize on the cooler portions of the wall. The tube is now sealed off from the pump and the drawn-down end painted black for a distance of a few centimeters. For the illumination I used a very simple modification of the 'light furnace' described in the earlier paper.

The iodine tube is fastened alongside of and in contact with a small Cooper-Hewitt mercury lamp (glass, not quartz). The bulb should project a centimeter or two beyond the cap on the positive electrode, and the drawn-down end should reach not quite down to the negative electrode bulb. Two small pads of thick asbestos paper should be placed between the two tubes, which are then securely fastened together with copper wire.

The Cooper-Hewitt lamp is supported in a clamp fastened close to the negative bulb, just beyond the end of the iodine tube, as shown in Fig. 1.

A cylindrical reflector is now prepared by cutting off the bottom of a beaker glass measuring about 12 x 25 cms., and silvering the outside with Brashear's solution. This can be done with a minimum amount of solution by rotating the beaker slowly in a glass or porcelain tray, tipped slightly on its longer side. A preliminary trial with water shows at once the minimum amount

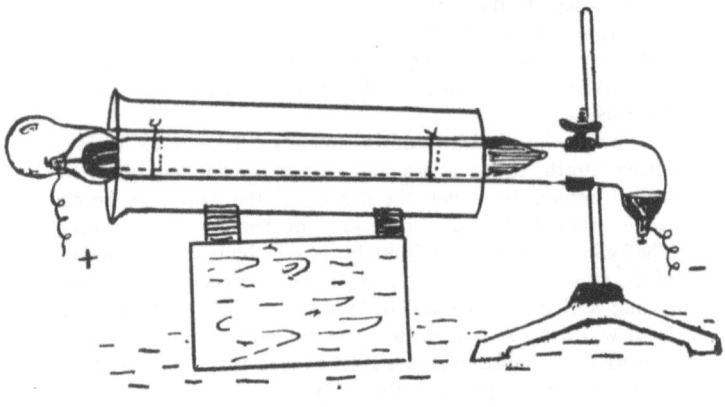

FIGURE 1

that can be used. It is sufficient if the solution wets one side of the beaker from one end to the other. After the silvering the inside of the beaker is cleaned with a cloth dampened with dilute nitric acid, and the hollow reflecting cylinder slipped over the iodine tube and mercury lamp. The lamp is clamped at a suitable angle for operating, say 5° from the horizontal, and started by tipping the clamp stand. The reflector should be supported so that the tubes are centrally placed. The luminous iodine vapor is viewed 'end-on' through the bulb. If a prism of about 8°, such as is used for mounting Lippman photographs, is placed in front of and close to, the bulb, it has the effect of bringing the tube into the horizontal position which is advantageous if an image is to be projected on the slit of a spectroscope.

This is the arrangement which I have used in all of the recent work, and besides having a very high efficiency, it is very easy to construct.

The fluorescence of the iodine is so intense that the doublets excited by the green mercury line can be seen widely separated in the first order spectrum of a large grating with a telescope of three meters focus. In a small prism spectroscope the complete resonance spectrum is extremely brilliant.

The excitation of the iodine vapor results from the circumstance that the green mercury line and the two yellow lines coincide with absorption lines of the iodine, of which, as I have previously shown, there are between forty and fifty thousand in the visible spectrum.

We will consider, first, the resonance spectrum excited by the green Hg. line. To obtain this pure, it is necessary to screen off the light of the two yellow lines. This can be done with a solution of neodymium chloride, or the double salt of neodymium and ammonium, and as the use of a fluid screen is impractical in the case of the method of illumination just described, it is necessary in this case to iluminate the tube with a large condensing lens. As a matter of fact, however, the yellow lines are comparatively feeble in the case of the Cooper-Hewitt glass lamp, and the doublets excited by the green line are so intense, that, in the greater part of the work, no screen has been employed.

The more recent investigations have brought some extremely interesting phenomena to light, especially with respect to the transfer of energy from the doublet series to the band spectra, as a result of the admixture of helium or other rare gases with the iodine.

On account of the complexity of the subject it will be necessary to touch briefly on some of the relations which have been discussed in the earlier papers.

The band absorption spectrum of iodine covers the spectrum range comprised between wave-lengths 5100 and 7700. It is made up of exceedingly fine lines averaging twenty to the Åugstrom unit in the green and yellow regions, or some 50,000 in all making the estimate on the above average. It is covered on the short wave-length side by a band of continuous absorption in the blue-green region, which makes the exact determination of its end impossible. In the red it has been followed by means of dicyanine plates sensitive to $\lambda = 9000$, and its termination discovered at about $\lambda = 7700$. A portion of this spectrum, in the vicinity of the green mercury line, reproduced from an earlier paper, is shown by Fig. a. Plate I. The entire spectrum, reproduced on the same scale, would be about eighty meters in length. In the yellow, orange and red regions the lines form fluted bands, or rather series of overlapping bands. In the green region there appears to be so much superposition of bands that all appearance of regularity vanishes. A good idea of the general appearance of this spectrum in the region 5460–5700 is given by Figure d. Plate I. This is in reality the emission spectrum of iodine in a vacuum tube, reproduced as a negative, and with the dispersion employed, could scarcely be distinguished from the absorption spectrum, except for the strong dark lines, which are iodine emission lines not belonging to the band emission.

The resonance doublets of orders 0, 1 and 3 excited by the green line of the Cooper-Hewitt mercury arc are also shown on this plate in their proper position (long lines).

With this as an introduction we will now take up the remarkable spectrum emitted by the iodine when illuminated by the green line of the Cooper-Hewitt lamp. This line is shown in coincidence with the absorption spectrum in Fig a. Plate I. As will be seen the main line falls nearly midway between two of the iodine absorption lines. It is probable that only the left hand absorption line is stimulated, as the width of the mercury line is not quite sufficient to enable it to reach the other. The short wave-length satellite is also in coincidence with an absorption line, but, for the present, we shall neglect the effect due to this. The resonance spectrum excited by the stimulation of this absorption line, consists of a series of close doublets (doublet separation about 1.50 ÅU.) very regularly spaced along the spectrum. For convenience we will designate, as before, the one in coincidence with the exciting line, as the doublet of 0 order, those lying on the long wave-length side as $+1, +2, +3$, etc., orders, and those on the other side as $-1, -2$, etc.

The doublet of 0 order is indicated on Fig. a. Plate I, immediately above the absorption spectrum. One line (5460.74) is in coincidence with the iodine absorption line covered by the mercury line, the other (5462.25) lies 1.5 ÅU. to the right. The former is in reality a re-emission of the absorbed energy without change of wave-length (Resonance radiation), and I have accordingly named this line the R. R. line. The other line we may designate the companion line. On the scale of Figure a the next doublet (+ 1 order) would lie on the right at a distance of nearly two meters.

By means of plates sensitized with dicyanine, which were prepared for me by Mr. Meggers, I have succeeded in photographing the doublets as far as the 27th order, with a large plane grating and a Cooke lens of one meter focus. This permits of wave-length determinations correct to about 0.1 ÅU.

This marks the end of the resonance spectrum, I believe, as the wave-length of the last doublet recorded on the plate was 7685, and the plates are highly sensitive to well beyond 8500. Moreover, the absorption spectrum terminates at about this point.

Photographs of the doublets are reproduced on Plate II, Figure j shows the orders 0, + 1, + 3, and + 4. The doublet of order + 2 is missing, though a pair of faint lines appear nearly in the position in which it should be found, (Figure e, Plate I) with long exposures.

Fig. b shows the doublets + 6 to + 13 inclusive, and Fig. c + 15 to + 22 inclusive; orders 9, 14, 16, 19 and 21 are also missing. The variable intensity of the doublets is also to be noted.

The law governing the spacing of these doublets will be discussed presently: for the moment we shall consider only the general nature of the phenomena.

Fig. d shows the doublets of order 0, + 1 and + 3 taken with a large plane grating and an objective of 3 meters focus (exposure 15 minutes), in superposition with the emission band spectrum of iodine electrically excited in a vacuum tube. All of the photographs, with the exception of Fig. a, are reproduced as negatives. The resolving power employed in the case of d was, of course, quite insufficient to completely resolve the band spectrum, as can be seen by comparing the width of the doublets with the width indicated in Fig. a. It, nevertheless, gives an idea of the relation of the doublets to the band absorption spectrum.

If we give a longer exposure we find that the doublets are accompanied by faint companion lines. These appear in Fig. e, which was exposed for an hour and a quarter. Some of these lines are due to the excitation of other iodine absorption lines by the satellites of the green mercury line, but others I feel sure result from the stimulation of the absorption line covered by the main line. The former come out strong when the iodine is excited by the quartz mercury arc, in which case the green line can be broadened until it covers all of the seven absorption lines between the two arrows in Fig. a.

If now we give a greatly prolonged exposure, we find that a band spectrum also appears. Fig. f is a 20-hour exposure for the same region of the spectrum. The doublets have fused to a wide band, owing to over exposure. The companion lines, above referred to, come out strong, and in addition there is a fluted band to the right of the doublets of order $+1$ and $+3$.

It will be observed that these doublets lie just within the heads of the fluted bands, a circumstance which is better shown by Fig. h, in which the heads of the bands are indicated by arrows. In the case of Fig. g the iodine tube, instead of being highly exhausted, contained Xenon at a pressure of 1.5 mm. As is apparent, the effect of the Xenon is to reduce tremendously the intensity of the doublets, and bring out strongly a number of fluted bands between the doublets, of which scarcely a trace can be seen in the case of iodine in vacuo. In the case of Fig. h, we have the iodine in helium at 4 mm. The doublets are still further reduced in intensity, the bands are stronger, and a new band appears at the center, no trace of which can be seen in g. The heads of the bands are not resolved, though on the original plate a number of the component lines can be seen to the left of the doublets. The doublet of the second order, which is missing, would fall at a considerable distance from the head of the band. There is in fact a group of lines at this point in Fig. f, but it is my opinion that they result from excitation of the vapor by some of the satellite lines; at all events none of them fits into the series of doublets excited by the main line.

If we compare Fig. h with Fig. d we see at once that the band spectrum emitted by iodine in helium, with monochromatic excitation is much simpler than the complete band spectrum. For example, there is in Fig. d a strong band-head at A, of which no trace appears in Fig. h. Moreover, fewer of the bands appear in the case of iodine in vacuo than in the case of iodine in helium.

If the excitation is by the quartz mercury arc the bands become more complicated, and in place of the doublets we have groups of lines, which will be discussed more in detail presently.

RELATION BETWEEN THE DOUBLETS AND THE BAND SPECTRUM

The absorption spectrum of iodine is made up of more or less regular fluted bands, resolvable under high dispersion into fine lines. The heads of these bands lie towards the region of shorter wave-lengths, and there is considerable overlapping which gives rise to considerable irregularity in appearance, especially in the green region. The emission spectrum of iodine, electrically excited in a vacuum tube closely resembles the absorption spectrum,

FIGURE 2

though they are not exactly complimentary, as will be shown in a following paper. Now the green line of mercury, which excites the series of doublets, lies just within the head of a well marked band in the emission spectrum, and it will be observed that the doublets of order $+1$ and $+3$ are similarly located. This was ascertained by superposing the resonance spectrum on a band emission spectrum. It is less well shown, except for the doublet of the $+3$ order by Figure d, Plate I, which was taken under conditions not well suited to emphasize the heads of the bands, the line spectrum being too prominent. The three bands above specified appear as emission bands accompanying the doublets, when the iodine is excited in vacuo, as shown diagrammatically by Figure 2, in which the doublets have been drawn a little longer than the lines forming the bands. The band accompanying the doublet of 0 order is not as strongly developed as the other two, and only its head shows in Plate I, Figure f.

By comparing the plates of the resonance spectrum with those of the band spectrum it has been found that the doublet of the fourth order also lies just within the head of a band. Above this point the relations have not yet been exactly determined, for the band spectrum accompanying the resonance doublets has not yet been photographed in the red. (Note added March, 1918: Prof. Okano and I have recently photographed the entire band spectrum in helium and have found that this relation holds at least in the case of the strong doublets up to the twenty-second order.) Though the fourth order doublet, which is faint, lies near the head of a band shown on the plate made of the electrically excited vapor, it does not occupy a corresponding position with respect to the band which forms a member of the simpler system shown in Figure h, the spacing of which is two-fifths of the distance between the doublets, *i. e.*, there are five bands between the doublets of first and third order.

It will be necessary to trace this simpler band spectrum throughout the orange and red region, before we can be sure that all of the strong doublets are located near the heads of the bands, and the missing ones near the tails.

The doublet of the sixth order is very strong, and it lies just within the head of a strong band shown by electrical excitation, and the same thing appears to be true of the eighth and tenth order doublets. The interesting point, however, is that a simple system of fluted bands, spaced apparently according to a law, similar to that which governs the spacing of the doublets, is excited by the stimulation of a single absorption line.

MULTIPLEX EXCITATION

If, instead of the glass Cooper-Hewitt lamp, we employ a quartz mercury arc (Westinghouse, Cooper-Hewitt) for the excitation of the iodine vapor, we find complicated groups of lines in place of the simple doublets. This is due to the fact that the green mercury line has broadened to such a degree that it covers a number of the iodine absorption lines. This we may call multiplex excitation.

The first point of interest which we should note is that the intensity distribution among the groups is practically the same as for the doublets, *i. e.*, groups of strong lines are built up around the strong doublets, weak groups around weak doublets, and only a few very faint lines at the points where the doublets are missing. This means that the dynamics of the vibrating system excited is very much the same in the case of the several absorption lines covered by the broadened mercury line.

The complexity of the groups depends upon the width of the green line which increases with the potential drop across the terminals of the quartz arc, as has been shown in previous communications.

If sufficient resistance is put in circuit with the arc to keep the potential down to 35 volts, the iodine emits the doublets only, Fig. j. Plate II. With the potential at 60 volts we have two new lines to the left of the doublets, as shown by Fig. k. Plate II, while with a potential difference of 110 volts we have the complicated groups shown by Figs. l and m, the latter showing the group of -1 order. Enlargements of groups -1, 0, $+1$, etc., are reproduced on Plate II, y and z, the upper with excitation by the arc at 120° volts, the lower at 60° volts. These groups are so similar in appearance, that, until very recently I have considered that the lines corresponded to each other, that is to say, that the fourth line from the left in each group was excited by the same absorption line. I now feel certain, however, that we must be a little careful about accepting this conclusion, from reasons which will appear presently.

In discussing the manner in which the groups are formed by multiplex excitation, we must recall that in the case of strictly monochromatic excitation, where a single absorption line only is stimulated, we have a series of doublets, the shorter wave-length component of the first doublet coinciding with the absorption line.

It has been found that the doublets conform very nearly to the following formula in which $1/\lambda$ represents the frequency of the left-hand component of the doublet of order m.

$$\frac{1}{\lambda} = 183075 - 3132\,m + m\frac{(m-1)}{2}13$$

or, putting it in words, that (approximately) the distance between the doublets increases by a constant amount as we pass from each one to the one of next higher order. The degree of accuracy with which this formula is followed will be discussed presently. The circumstance that we have a group of lines formed around the (unresolved) absorption lines which are excited by the broadened mercury line, furnishes us the clue as to how the groups originate. As was stated in the previous paper:

"These groups originate in the following way: The seven absorption lines which are covered by the broadened green mercury line are simultaneously excited, and the vapor emits these seven wavelengths without change. These lines we may call the R. R. lines (resonance radiation). Each one of these is moreover the first member of a series such as is expressed by the formula previously given. The R. R. lines are not resolved by the spectrograph employed in photographing the resonance spectra and consequently appear superposed. But each one is accompanied by one or more companion lines, lying to the right or left, and it is these companion lines which form the group of 0 order.

The actual width of the group of seven R. R. lines is only about 1/30 of the width of the group formed by the companion lines."

Let us now see *how* the groups of higher order are built up. Suppose each of the seven R. R. lines to be the first member of a series such as was represented by our formula and suppose that for each one we have the same values of the constants. Suppose moreover that each member of any given series is accompanied by a companion line. In this case the group of 0 order will be exactly duplicated at intervals along the spectrum. The center of each group will be composed of seven superposed lines (in reality separated by the same small intervals as the R. R. lines) each one of which is accompanied by a companion line to the right or left as the case may be. In an earlier paper I spoke of the seven superposed lines as the 'core' of the group. As a matter of fact, the spacing is not exactly the same for the seven series of main lines, consequently, as we ascend to higher group orders they begin to separate, even with the resolving power employed in photographing the resonance spectra. This accounts for the fact that the groups of higher order differ in appearance from those of lower.

Resonance Spectra of Iodine

I have photographed the groups of 0 and + 1 order with the seven-inch grating and 3 meter objective in the fourth order spectrum with an exposure of 48 hours. The lines were very faint but perfectly sharp. The appearance of the two groups is shown in Fig. 3. The resolving power in this case was but little less than that required to separate the iodine absorption lines and we find that the center of the group of 0 order is a narrow band (line No. 6 in the figure) made up of five barely resolved lines. It was possible to count the lines by holding the plate somewhat foreshortened under a magnifying glass. These are the R. R. lines. The other lines which form the group are the companion lines, and the fact that there are more of them than R. R. lines suggests that

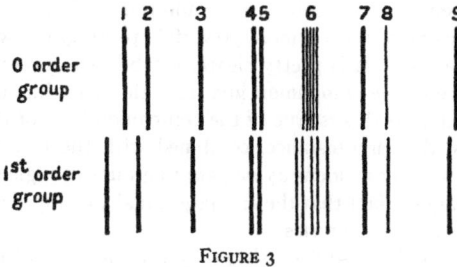

FIGURE 3

probably some of the R. R. lines have two companions instead of one.

Passing now to the first order group, we find that the main lines which form the core (each one of which belongs to a series of which a R. R. line is the first member) are more widely separated than in the 0 order group, the spectral range having about doubled. This is due to the fact that the value of the constant in the second term of our formula is not the same for each series.

We will now consider the subject of the companion lines. In the case of the doublets excited by the Cooper-Hewitt lamp the companion lines lie on the long wave-length side of the main lines, at long distances which gradually increase with increasing group order. The widths of the doublets in the various orders are given in the following table:

Order	Width	Order	Width
0	1.48	15	2.04
1	1.54	17	2.19
3	1.64	18	2.25
5	1.72	20	2.43
8	1.76	22	2.45
10	1.85	23	2.53
11	1.90	25	2.50
12	1.95	27	2.80
13	1.96		

The increment is not quite regular and it is my hope that a new set of plates made with a more powerful spectrograph will show no discrepancies. It is pretty clearly established, however, that the distance of the companion line from the main line increases progressively. If this is true of the companion lines of the other main lines, this circumstance, combined with the fact that the group of main lines widens as we pass to groups of higher orders, explains fully the fact that the groups gradually change in appearance as we ascend the series.

Groups 4, 5, 6, 7 and 8 are shown by Figs. n and o of Plate 11, the former excited by the Cooper-Hewitt lamp, the latter by the quartz arc at 115 volts. In the case of Fig. n we have faint series excited by the two yellow mercury lines, one of which (5790) lies within the fifth order group excited by the green line. The lines marked by small crosses are the ghosts of the yellow mercury lines, and should not be confused with the resonance lines.

In Fig. o it will be observed that the doublets shown in Fig. n have become relatively weak, and that we have a new series of strong doublets displaced towards the left with respect to the old ones. This is due to the fact that, in the case of the quartz arc operating at 115 volts, the green mercury line is strongly reversed and the excitation of absorption line No. 3 (Plate I, Figure a) becomes relatively weak, as it coincides with the reversed core of the mercury line. In the seventh order of Fig. n there are four lines, the first and third form the doublet excited by the green mercury line, the other two belong to a series of doublets excited by the yellow line 5790. The same condition is found in the fourth order group, the dotted doublet in this case lying to the right of the doublet of order −1 excited by 5790. Similar complications, of course, occur at other higher orders.

If we could excite the iodine absorption lines one at a time there would be no difficulty in finding out how the groups are built up, but this is impossible with present facilities.

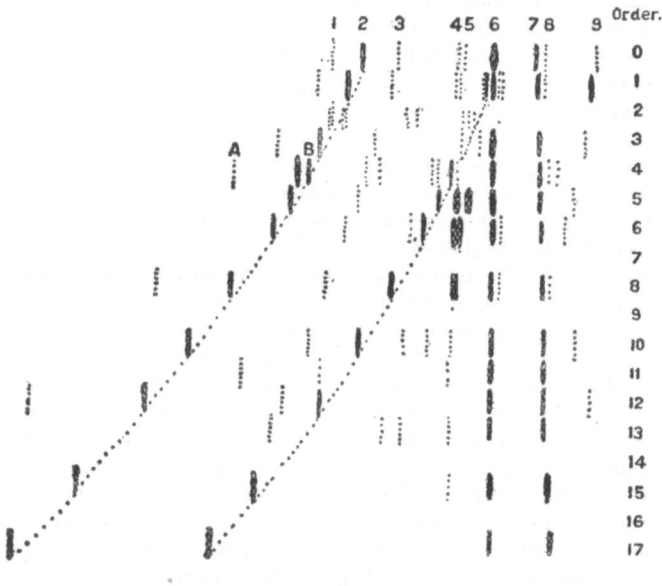

FIGURE 4

By varying the voltage at which the mercury lamp operates, and by filtering the light through bromine vapor, some clues have obtained regarding the relations existing between the absorption lines and the lines forming the groups, but a complete analysis has not yet been made.

In Figure 4, I have given a diagram of the groups up to the seventeenth order excited by the green mercury line of the quartz lamp operating at 115 volts. The doublets (lines 6 and 7) excited by the Cooper-Hewitt lamp, appear in all of these groups, though they are relatively faint, owing to the reversal of the exciting line, and these doublets are brought into coincidence in the diagram.

When the iodine is excited by the lamp operating at 60 volts, lines 2, 4, 5, 6 and 7 appear, in the group of 0 order, line 6 being of course the unresolved complex of emission lines corresponding to the absorption lines covered by the green mercury line. Line 7 is the companion line which together with the 'R. R.' line corresponding to absorption line 3, forms the doublet of 0 order. The doublets of higher order lie immediately below, the increasing distance between the components being very apparent.

Now line 2 is a companion line to the R. R. line corresponding to absorption line 4, indicated also by line 6 in the diagram. These two lines form another doublet of zero order. The higher orders do not lie immediately below, but drift to the left, as indicated by the dotted lines. This is due to the fact that the constant in the second term of the formula is a little less than in the case of the first series of doublets considered, in other words the doublets are closer together.

In the group of the first order the main line of this series of doublets can be separated from the main line of the other series only in the fourth order spectrum of the grating. In the third order group it is so far detached, that it was confused for a long time with line 5 of the first order group. If we compare the orders 0 and 6 we shall see another case of this kind: If it were not for this diagram arrangement of the groups, we should probably assume that the first line to the left of group 6 corresponded to line 1 in group 0, whereas the diagram shows clearly that it corresponds to line 2. Moreover, it appears in the 60-volt excitation, which does not bring out lines 1 and 3.

In the construction of the diagram it is, of course, necessary to leave blank spaces for the missing orders, otherwise the corresponding lines will not lie on a smooth curve.

It is a little difficult to explain in words just how this diagram is to be interpreted, though it is clear enough if the theory of the group formation which I have given is understood. All of the lines with the exception of 6 in the 0 order group must be companion lines, line 6 being made up of the unresolved R. R. lines.

In the case of the doublets, the superposition of which form the other groups we must distinguish between what I have called the main line and the companion. As we run up the diagram the main lines should lie on curves intersecting line 6, for example, the dotted curve shown which belongs to the 2, 6 doublet.

I have not yet been able to identify certainly any other main lines, though I suspect that the one corresponding to companion line 9, descends from line 6 on a curve sloping to the left at a lesser angle than the dotted curves, *i. e.*, at about the angle taken by companion line 3.

Various modifications in the conditions of excitation have been made with a view of establishing which absorption lines are responsible for the various doublets.

For example, it was found that the lateral emission and the end-on emission of a Cooper-Hewitt lamp showed a very different intensity distribution in the green mercury line, as shown by Figures r and s, Plate II, which were made with a very fine plane grating by Dr. Anderson. The same are shown in coincidence with the absorption spectrum of iodine (reproduced as negatives) by the small circular prints on Plate II, v. If the iodine vapor is excited by the lateral emission of the lamp, as with the 'light-furnace', companion line No. 1 appears in addition to the strong doublets. See 0 and + 1 orders of Figure j, Plate II. After several failures I succeeded in obtaining a record of the iodine resonance excited by the end-on emission, and in this spectrum companion line No. 2 appeared also. Now companion line No. 1 does not appear in the case of excitation by the quartz arc operating at thirty-five volts, and the short wave-length satellite of the green line is weaker, with respect to the main line, in this case, than in the case of the Cooper-Hewitt lamp, as is shown by Figures t and u, Plate II (t being the Cooper-Hewitt line and u the quartz arc). This makes it appear probable that companion line No. 1 arises from the excitation of the absorption line which is in coincidence with the short wave-length satellite.

Companion line No. 2 is probably due to the excitation of absorption line No. 4. It comes out with excitation by the 'end-on' emission of the Cooper-Hewitt lamp owing to the broadening of the main line which occurs under this condition, and for the same reason it is the first line to appear when the terminal voltage of the quartz arc is increased. No very definite conclusions have been drawn from the numerous experiments which have been made with the exciting light filtered through bromine vapor and nitrogen tetroxide. With a potential of 90 volts on the quartz arc companion lines 4 and 5 appear. If the exciting light is filtered through bromine vapor contained in an exhausted bulb about 30 cms. in diameter, line No. 5 disappears in the groups of order 0 and +1. In the third order group line No. 5 is much stronger than 4 and bromine filtration of the exciting light equalizes the intensity. Line No. 4 must, therefore, be due to the excitation of an absorption line which is not in coincidence with a bromine line, and which is first covered by the mercury line when the lamp operates at 90 volts. This seems to be absorption line No. 5, while the other component, which is removed by filtration of the exciting light through bromine, is probably due to absorption line 6.

With a potential of 110 volts on the lamp, companion line No. 3 appears, and this also is removed by the bromine filtration of the exciting light, as is shown by Figures p and q, Plate II, in which p is the resonance spectrum obtained when the exciting light is filtered through bromine. It appears to be due to the stimulation of absorption line 7 which is in coincidence with a bromine line.

The difficulty in interpreting the results obtained is due to the fact that the mercury line widens both to the right and left as the voltage increases, so that two absorption lines may be attacked simultaneously. If this happens, we can differentiate between them only if one of them is in coincidence with a bromine line and the other not. What is most needed just now is one or more other filters similar to bromine vapor, but I have not been able to find anything with sufficiently narrow lines, though I have tried a number of vapors which looked promising. What would be still better would be to alter the wave-length of a narrow exciting line so as to cause it to pass by degrees from one absorption line to the next.

EXCITATION BY THE YELLOW LINES

The resonance spectra excited by the two yellow lines have not been completely investigated as yet, though a large number of photographs have been made. Each yellow line excites a series of nearly equidistant groups which resemble roughly the groups excited by the green line. Six pairs of these groups, from − 1 order to + 4 order, photographed with rather low dispersion are shown by Figure i, Plate I. Enlargements of the groups excited by the line 5791 and the line 5769 of the quartz arc at 135 volts are reproduced on Plate I, w and x. In this case the excitation was by the quartz mercury arc operating at 140 volts, the green line having been cut off by means of a glass trough filled with a solution of eosine. Some difficulty was found in securing the spectrum excited by the Cooper-Hewitt arc, as the yellow lines are comparatively weak in this case, but satisfactory results were finally obtained with the light furnace, the iodine tube being wrapped around with a sheet of gelatine stained to a deep orange yellow.

In this case each yellow line excited a series of doublets, but both series were much more irregular than the series excited by the green line.

The separation of the components of the doublets excited by the 5790.7 line varied in an irregular manner from 2.1 to 5.6 ÅU. In the case of the excitation by the 5769.6 line we have also a series of doublets, though the companion line is missing at the zero order, in other words the R. R. line has no companion. The separation of the components of the doublets is less irregular in this case, varying from 4.8 to 5.4 ÅU. The table of wave-lengths will be given in the following paper.

No. 4

The Series of Resonance Spectra

(In collaboration with M. Kimura)

In the previous communication a general account of the results which have been obtained, up to the present time, on the resonance spectra of iodine has been given.

The present paper will deal with the measurements of wavelength of the lines in the groups, and the subject of the series law which governs their spacing.

The wave-lengths of the lines in the groups of 0 and + 1 order were determined from plates made in the fourth order spectrum of a large plane grating with a telescope of 3 meters focus. They are correct probably to 0.01 ÅU. The groups + 2, + 3, and + 4 were made in the second order spectrum, and the higher order groups in the first order spectrum.

The series which has been most definitely determined, and to which the greatest amount of study has been given is the series of strong doublets excited by the Cooper-Hewitt lamp.

The two components of each doublet appear to be of equal intensity, although, in the case of two or three, a different ratio appears in the photograph as a result of absorption. It was found, as has been stated in earlier papers, that the first order group, which is usually recorded with the component of shorter wave-length three or four times as intense as the other, comes out with its lines of nearly equal intensity if the lateral branch of the iodine tube is cooled to zero, while the right hand component disappears entirely if the light from the tube is passed through a large glass bulb containing iodine vapor, before it enters the spectroscope.

In studying the series law it has been found necessary to reduce all wave-length to vacuum, and convert them into frequencies.

We will take up first the study of the doublets, the wavelengths of which and their reciprocals are given in the following table, on the International Scale and reduced to vacuum.

Series of Resonance Spectra

DOUBLETS EXCITED BY GREEN LINE OF COOPER-HEWITT LAMP

Group Order		$\frac{1}{\lambda}$ (Obs.)	Freq. Dif.	$\frac{1}{\lambda}$ (Cal.)	Difference between Obs.&Cal.
0	5462.23 (Hg. or RR Line) 5463.74	183075 183025	50	183075	0
1	5526.55 5528.10	180945 180894	51	180942	+3
2	Missing				
3	5658.71 5660.38	176719 176667	52	176715	+4
4	5726.59 5728.25	174624 174573	51	174621	+3
5	5795.79 5797.51	172539 172488	51	172539	0
6	5866.14 5867.85	170469 170420	49	170470	−1
7	Missing				
8	6010.66 6012.50	166371 166320	51	166370	+1
9	Missing				
10	6160.63 6162.48	162321 162272	49	166322	−1
11	6237.68 6239.56	160316 160268	48	160316	0
12	6216.16 6218.14	158324 158270	50	158324	0
13	6396.08 6398.05	156346 156297	49	156344	+2
14	Missing				

Group Order		$\frac{1}{\lambda}$ (Obs.)	Freq. Dif.	$\frac{1}{\lambda}$ (Cal.)	Difference between Obs.&Cal.
15	6560.56 6562.68	152426 152377	49	152423	+3
16	6645.0 6647.0	150489 150443	46	150481	+8
17	6731.12 6733.28	148564 148516	48	148552	+12

From this point on values determined from plates made with telescope of 1 meter focus. They are correct only to about 0.1 ÅU.

Group Order		$\frac{1}{\lambda}$ (Obs.)	Freq. Dif.	$\frac{1}{\lambda}$ (Cal.)	Difference between Obs.&Cal.
18	6818.63 6820.91	146657 146608	49	146636	+21
19	Faint and masked by mercury line				
20	6998.96 7001.39	142878 142828	50	142842	+36
21	Missing				
22	7186.23 7188.68	139155 139107	48	139099	+56
23	7282.39 7284.92	137318 137270	48	137247	+71
24	Missing				
25	7480.4 7482.9	133682 133638	44	133580	+102
26	Missing				
27	7685.7 7688.5	130110 130060	50	129964	+146

The first point established by this table is that, while the separation of the components of the doublets increases progressively from 1.51 °AU. at 0 order to 2.5 ÅU. at the twenty-seventh order, *the frequency difference between the components is a constant; 50.* The extreme low values 46, and 44 found in the sixteenth and twenty-fifth order are undoubtedly due to the fact that the lines were extremely faint, and the wave-lengths could not be very accurately determined. The last doublet (the twenty-seventh order) was fairly strong, and the frequency difference in this case is exactly the same as in the case of the 0 order.

We will now consider the law governing the spacing of the doublets along the spectrum, applying the calculations to the first member of each doublet (shorter λ component). If we confine our attention to the first few orders, it seems as if the distance between the doublets increased by a constant small increment. This would mean a constant second difference of wavelengths. It was found, however, that this condition held only for the first few orders. The reciprocals of the wave-lengths were next examined, and it was found that a constant second difference existed, at least over a considerable range of the spectrum.

If this condition held rigorously the series would be represented by the formula,

$$\frac{1}{\lambda_m} = 183075 - 2132\,m + 13\frac{m(m-1)}{2}$$

in which λ_m is the wave-length of the doublet of the mth order, 2132 is the frequency difference between orders 0 and $+$ 1, 13 the constant second difference of frequency, and m the order of the doublet. The most accurate value of the second constant would be obtained by calculating it from a doublet of high order, as a small error would be enormously magnified by the term $\frac{m(m-1)}{2}$ the value of which is 351 for the twenty-seventh order.

Calculating the constants 2130 and 12.2 from orders 0 and 5 gave calculated values of $1/\lambda$ which differed from the observed by the following amounts:

Doublet Order	Difference	Doublet Order	Difference
1	0	7	+ 2
3	+ 1	10	+ 4
4	− 1	13	+ 14
5	− 1	18	+ 65
5	0	23	+ 166
6	− 1	27	+ 284

The large discrepancies in the higher orders are due to incorrect determination of the constants. In spite of this though the series is well represented up to the seventh order. The following formula gave the best results over the entire range:

$$\frac{1}{\lambda_m} = 183075 - 2131.414 = m - 12.734 \frac{m(m-1)}{2}$$

The values given in the table were calculated by this formula and, as will be seen, the agreement is good up to the doublet of fifteenth order. There is a small discrepancy in the orders 1, 3 and 4, which appears to be inevitable if the constants are so chosen as to make the formula cover a wide range. Of course, the formula is not correct for the entire series, and though we have tried formulae involving higher powers of m than the square, we have been unable to develop anything superior to the one given. The discovery of the fact that the frequency difference between the components of the doublets is a constant, has been of assistance in picking out other series of doublets in the series of complicated groups excited by the quartz arc.

For example, we may take the wider doublets shown united by dotted lines in the previous paper (Fig. 4).

The frequency differences for these doublets are given in the following table:

Order	Freq. Dif.	Order	Freq. Dif.
0	161	8	158
1	159	10	157
3	159	12	157
5	158	15	152
6	157	17	156

It will be remembered that the frequency difference of the first series of doublets considered was 50.

The spacing of this series along the spectrum is only fairly well represented by the formula:

$$\frac{1}{\lambda_m} = 183075 - 2119\,m + 13\,\frac{m(m-1)}{2}$$

The observed and calculated values of $1/\lambda$ for the components of longer wave-length (the companion line is to the left in this case) are given in the following table. It will be observed that the doublets are missing in the fourth, eleventh and thirteenth orders, as well as in the orders in which the doublets of the first series failed to appear:

Order	$\frac{1}{\lambda}$ (Obs.)	$\frac{1}{\lambda}$ (Cal.)	Dif.
0	183075	183075	0
1	180956	180956	0
3	176754	176757	3
5	172599	172600	1
6	170543	170556	13
8	166470	166487	17
10	162448	162470	22
12	158479	158505	29
15	152621	152655	33
17	148786	148820	34

In the following table are given the wave-lengths and their reciprocals, on the International Scale and reduced to vacuum, of all of the lines in the groups between 0 and 17, in the case of iodine vapor excited by the quartz mercury arc operating at 115 volts.

The doublets excited by the Cooper-Hewitt lamp are marked thus * and the other doublets which we have studied thus † This table corresponds to the diagram in the previous paper. In the fourth order group lines A and B were added from an old series of measurements. The line between them is the only one which appears on our recent plates, and this line only is given in the table:

No. of Line	λ (O Order)	$\frac{1}{\lambda}$	No. of Line	λ (Second Order)	$\frac{1}{\lambda}$
1	5			5585.08	179048
†2	5457.43	183236		86.33	179008
3	5458.33	183189		86.81	178992
4	5460.88	183121		89.05	921
5	5461.07	183114		89.42	909
†*6	5462.23	183075		91.02	858
*7	5463.74	183025		91.17	853
8	5464.05	183014		91.38	847
9	5466.04	182948		93.33	784
	First Order			93.77	770
1	5520.30	181149			
2	†5521.33	181115		*Third Order*	
3	5522.92	181064			
4	5525.17	180989	1	5651.06	176958
5	5525.38	180983	2	†5652.49	176913
	†5526.20	180956	3	5654.55	176849
	.22	955		†5657.57	176754
6	.47	47	6	*5658.71	176719
	*5526.55	45	7	*5660.38	176667
	.71	39	9	5661.97	176617
	.80	36		5663.15	176580
7	*5528.10	180894			
8	5528.39	180884			
9	5530.06	180830			

and in addition faint lines as follows:

λ	λ
5656.87	5658.96
57.17	59.67
57.35	59.50
57.98	60.70
58.24	61.02

Fourth Order			Eighth Order		
No. of Line	λ	$\frac{1}{\lambda}$	No of Line	λ	$\frac{1}{\lambda}$
	†?5719.62	174837		5998.6	166704
	5722.05	174763		†6001.38	628
	22.55	4747		04.88	531
	24.47	4688		†6007.07	471
	24.72	4681		09.32	411
	†5725.13	4668		09.40	406
	25.35	4661		*6010.66	371
	*5726.59	4624		10.91	364
	26.84	4616		*6012.50	320
	*5728.25	4573		12.80	310
	28.56	4564			
	28.95	4552			

Fifth Order			Tenth Order		
	†5788.45	172757		†6149.87	162605
	91.03	2681		54.07	494
	†5793.77	2599		†6155.81	448
	94.48	2578		57.52	403
	94.88	2566		58.31	382
	*5795.79	2539		59.14	360
	*5797.51	2488		*6160.63	321
	98.70	2452		*6162.48	272
				63.65	241

Sixth Order			Eleventh Order		
	†5858.23	170700			
	60.8	625			
	63.2	555		6228.79	160545
	†5863.6	543		31.44	477
	64.75	510		36.12	356
	64.9	505		*6237.68	316
	*5866.14	469		*6239.56	268
	*5867.85	420		42.0	205
	68.1	413			
	68.1	395			

Twelfth Order			Fifteenth Order		
No. of Line	λ	$\frac{1}{\lambda}$	No. of Line	λ	$\frac{1}{\lambda}$
	6294.87	158859		†6545.65	152773
	99.40	745		†6552.19	621
	†6303.71	636		59.05	461
	08.57	514		*6560.56	426
	†6309.99	479		*6562.64	378
	*6316.16	324			
	*6318.14	274	Seventeenth Order		
	19.6	237		†6714.00	148923
Thirteenth Order				†6721.03	787
				*6731.09	564
	6388.24	156538		*6733.27	516
	92.14	442			
	92.82	425			
	94.54	383			
	*6396.08	346			
	*6398.05	297			

EXCITATION BY THE YELLOW LINES

We have measured the wave-lengths of the lines in the resonance spectrum excited by the yellow lines of the Cooper-Hewitt arc, and the quartz arc operating at 115 volts. The values given in the following table are on the International Scale and reduced to vacuum. They were determined from plates made with the plane grating and Cooke lens of 1 meter focus, and can be considered correct only to about 0.1 ÅU. We have, however, made some measurements of the doublets photographed with the 3-meter lens, which are correct probably to 0.02 ÅU. and as the same irregularities were found in the spacing, we have not thought it worth while to measure the complete spectrum to the highest degree of accuracy.

The wave-lengths are given in the following table. The lines or doublets excited by the Cooper-Hewitt lamp are marked thus *:

Series of Resonance Spectra

EXCITATION BY HG. 5769.6 (5771.2 REDUCED TO VACUUM)

	$\dfrac{1}{\lambda}$		$\dfrac{1}{\lambda}$
− 1 Order		Fifth Order	
5701.8	175383	6139.0	162893
5705.2	278	41.0	839
		*43.1	784
0 Order		Sixth Order	
5766.0	173430		
67.8	376	6215.1	160898
*5771.2	274	17.0	849
75.2	154	18.9	800
		*22.1	717 } 129
+ 1 Order		*27.1	588
5834.4	171397		
38.5	277	Seventh Order	
*5842.5	159 } 137	6294.9	158858
*5847.2	022	99.6	740
		*6303.5	641
Second Order			
5911.2	169170	Eighth Order	
14.7	070		
*5915.8	039 } 143	6374.3	156879
*5920.2	168896		
Third Order		Ninth Order	
5985.2	167078	6459.7	154806
88.3	166992	63.7	710
*90.0	945 } 142	*67.9	609 } 126
*95.1	803	*73.2	483
Fourth Order		Tenth Order	
6060.4	165006	6544.1	152809
64.6	164891	48.7	702
69.1	164769	*53.0	602

The series excited by the Cooper-Hewitt lamp in the case of the 5769 line differs from that excited by the green line in a number of respects.

In the first place at the point of excitation we have only the R. R. line with no companion. At orders 1, 2, 3, 6 and 9 we have doublets, 4 and 8 are missing, and at 5, 7 and 10 we have single lines.

The $1/\lambda$ difference, in the case of the components of the doublets is not constant, as in the previous case, but varies from 143 to 126.

As to the spacing of the doublets along the spectrum we find that in this case the $1/\lambda$ difference is very nearly *constant*, as is shown by the following table:

Order	$\dfrac{1}{\lambda}$	$\dfrac{1}{\lambda}$ Difference	Order	$\dfrac{1}{\lambda}$	$\dfrac{1}{\lambda}$ Difference
−1	175383		5	162784	
0	173274	2109	6	160715	2069
1	171159	2015	7	158641	2074
2	169039	2120	8		
3	166945	2094	9	154612	2014
4		2080	10	152602	2010

The variation is irregular, and it is obvious that the series is of a different type from the one excited by the green line, a portion of which at least was well represented by a formula.

In the case of the excitation by the 5790.7 line, we obtained different values of λ in the case of the Cooper-Hewitt lamp, consequently these values only are given in the table. It is probable that in the case of the quartz lamp at 115 volts reversal of the line causes the disappearance of the doublets excited by the lamp running at a lower temperature:

EXCITATION BY COOPER-HEWITT 5790.7 (5792.3 RED. TO VAC.)

− 2 Order			Fourth Order		
λ	$\frac{1}{\lambda}$	$\frac{1}{\lambda}$ Dif.	λ	$\frac{1}{\lambda}$	$\frac{1}{\lambda}$ Dif.
5658.6 5660.3	176722 669	53	6084.3 88.3	164357 249	108
− 1 Order			Fifth Order		
5722.1 5723.2	174761 727	34	6163.9 66.1	162235 177	58
0 Order			Sixth Order		
5792.3 5797.9	172644 476	168	6242.3 45.0	160197 128	69
First Order			Seventh Order		
5871.3 5873.4	170320 259	61	6325.8 29.2	158083 157997	86
Second Order			Eighth Order		
5936.8 38.9	168441 380	61	6404.0 06.7	156152 086	66
Third Order					
6010.8 13.3	166367 298	69			

In the case of the series excited by this line the doublets are present in all orders, but the $1/\lambda$ difference between their components varies in a very irregular manner from a minimum value of 34 to a maximum value of 168.

The spacing of the series along the spectrum is more regular, however, the $1/\lambda$ differences being as follows (the last significant figure is omitted):

196	205
211	211
232	193
188	212
207	200
203	209
209	

It does not appear to be worth while at this stage of the investigation to give the wave-lengths of the lines in the more complicated groups excited by the quartz arc operating a various voltages, as the simpler series excited by the Cooper-Hewitt lamp does not appear at the present time to conform to any law. The cause of this may appear when the relation of this spectrum to the band spectrum developed when the iodine is in helium, has been studied. This will require exposures of many days, however.

No. 5
Band and Line Spectra of Iodine

(In collaboration with M. Kimura)

Iodine vapor is of peculiar interest spectroscopically, in that it is one of the few substances which can be caused to emit line spectra of an almost infinite variety, of definite types, and very regular structure, by excitation with monochromatic light, as has been shown by one of us. These resonance spectra are very intimately associated with the complicated banded absorption and emission spectra, on which account it has seemed very desirable to make a comprehensive study of the spectra of this element excited by other means under the highest dispersion possible.

The present paper will deal chiefly with the electrical excitation of the vapor in vacuum tubes. In the course of the investigation we made the discovery that many of the lines of the line spectrum are complex under high dispersion, appearing as doublets, triplets, quadruplets, and quintuplets, the total width of the group in the case of the quintuplets being about 0.35 A. These complex lines behave in a most remarkable manner in the magnetic field, and we made a very exhaustive study of the Zeeman effect which they exhibit. This subject will be taken up in a subsequent paper.

The different types of spectra emitted by iodine vapor under different conditions of excitation were studied by Konen[1] nearly twenty years ago. In vacuum tubes excited electrically he found a band spectrum and a line spectrum, the relative intensities of which depended upon the diameters of the tubes, current-densities, vapor-densities, and other circumstances. Konen gives the wave-lengths of about 350 lines in the range of spectrum from λ 3030 to λ 5787, but makes very little mention of the band spectrum, stating that it was so feeble that it could be photographed only with a direct-vision prism-spectrograph (exposure eight to

[1] *Annalen der Physik*, 65, 257, 1898.

nine hours) which gave a spectrum less than 3 cm. in length for the range 5700–3000. We have, however, so improved the conditions of excitation that we have been able to photograph this band spectrum in the fifth-order spectrum of a large plane grating, with an objective of three meters focus—that is to say, with apparatus capable of completely resolving the absorption spectrum, which, as has been shown by one of us, contains in the neighborhood of 40,000 lines in the visible region. The spectrograms from which Konen's measurements of wave-lengths in the line spectrum were made were obtained with a small concave grating of one meter radius in the first-order spectrum, and he gives 0.04 Å. as his mean error. We are, however, in agreement with Kayser, who considers the limit of accuracy to be more nearly 0.1 A. Konen was unable to secure photographs in the second order on account of the faintness of the light. We have, of course, had no difficulty in photographing this spectrum in the fifth order, as the lines can be made very much brighter than the bands.

We found, in the early stages of the work that the insertion of a spark-gap or capacity in the circuit increased enormously the intensity of most of the lines and suppressed almost completely the band spectrum. There were, however, other lines which were reduced in intensity, and a few which showed little or no change. This circumstance has been mentioned briefly by Goldstein, and Stark has also alluded to it, classifying the lines which were increased in intensity as 'spark lines', the others as 'arc lines'.

In the preliminary part of the work we had considerable difficulty in finding suitable electrodes. Platinum is very rapidly attacked by the ionized iodine vapor, and deposits in the form of a brownish coating of very low reflecting power, which is probably a compound of the metal with iodine. We finally adopted tubes provided with external electrodes of tin foil. These tubes were of the form shown in Fig. 1. The bulbs were about 4 cm. in diameter and 15 cm. in length, joined by a capillary, which was blown out in the form of a thin bulb at A, for the emergence of the light. The process of exhaustion was as follows: A few flakes of iodine were introduced into the bulb through the tube B, which was then sealed. The flakes were then brought to the bottom of the tube B, and the tube C put in communication, through a U-tube immersed in liquid air, with a Gaede pump. If liquid air is not available, a tube filled with fragments of caustic potash should be introduced between the tube and the pump, to hold back the iodine vapor. During the process of exhaustion the bulbs must be strongly heated with a Bunsen flame. Before the tube is sealed off from the pump a small flame should be applied cautiously to the bottom of the tube B, until the iodine has entirely sublimed to the upper part of the tube. It is also a good plan to test the vacuum in the following way: Wrap the bulbs with tin-foil electrodes and start the discharge, using a coil capable of giving a six- or eight-

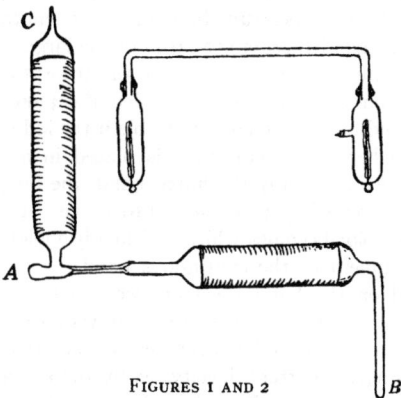

FIGURES 1 AND 2

inch spark; then touch the walls of the bulbs with cotton wet with liquid air. If the capillary is very fine, the vacuum in the bulb nearest the pump will usually be found to be much higher than in the second. At very low pressures the color of the discharge in iodine vapor is chamois-yellow, and the exhaustion should continue until this condition obtains in *both* bulbs, when they are cooled with liquid air or solid CO_2. If neither of these substances is available, immerse the tube B in a mixture of ice and salt. A yellow discharge in one bulb and a pink discharge in the other indicate that nitrogen is still present in the bulb beyond the capillary.

In our experiments we found that, to get the line spectrum at its brightest, the diameter of the capillary should be not over

0.15 mm., and considerable practice was required before suitable tubes could be produced. They were drawn down from 6 mm. tubing, which was first heated until the walls nearly collapsed. In the latter part of the investigation we found that very satisfactory iodine tubes could be made by using electrodes of thin platinum foil enameled with a thin layer of soft glass, which was smeared on in the flame of a blast lamp. These electrodes last a long time, heavy currents can be used, and the capillary can be a millimeter or more in diameter.

The spectrum of the electrically excited iodine vapor is made up of a fluted band spectrum between wave-lengths 5200 and 7000, which in the fifth-order spectrum of the large plane grating shows a structure comparable to that of the absorption spectrum, and a continuous band between wave-lengths 4300 and 4800 Å. This latter band becomes relatively feeble if the iodine is at a low pressure, as when the lateral tube is placed in a refrigerating medium, and we have only the fluted band, the integrated color of which is the chamois-yellow referred to above. As the pressure increases, the color becomes white and finally violet-blue, owing to the development of the continuous band.

Superposed on the band spectrum we usually have the line spectrum also, though it is nearly absent in wide tubes with small current-densities. It may be developed by constricting the tube or by increasing the current-density, as by the introduction of a condenser in the circuit. With the tubes provided with external electrodes it can be brought out strongly by means of a spark-gap placed in parallel with the tube. It may also be brought out, as we found, by merely heating the discharge tube to a high temperature by means of a burner. This indicates that it probably results from the dissociation of the iodine molecule. As the line spectrum develops in intensity, the band spectra fade away and finally disappear. Even in the tubes provided with external electrodes we found that traces of CO appeared after prolonged use. A yellow discoloration of the glass also developed, and it seems probable that this gas or CO_2 is liberated from the glass by the action of the ionized iodine.

A portion of the band spectrum with the lines superposed and the line spectrum alone are reproduced as negatives in Plate III, *a*. The latter was taken with a condenser in parallel (internal electrodes). The influence of the current-density is well shown by d and e, which reproduce the greater part of the visible spectrum. The lower portion of each was taken without spark, the upper with spark in parallel with the tube (external electrodes). For example, line 5234 is strong in the upper spectrum of a and the lower spectrum of d, while it has disappeared entirely in the lower spectrum of a and is relatively weak in the upper spectrum of d. Line 5245 behaves in exactly the reverse manner.

The disappearance of the band spectrum and the appearance of the lines can be brought about gradually by means of a variable capacity. We are of the opinion that it is the result of dissociation resulting from elevation of temperature, for we have succeeded in bringing about the same change by heating the narrow part of the tube with a Bunsen burner. The tube (Fig. 2) used in this experiment was provided with internal electrodes sealed into glass bulbs, which we joined by a quartz tube having a bore of about 3 mm., and cemented into the bulbs with sealing-wax. This tube, when excited by the coil, gave the band spectrum, with scarcely a trace of the lines (Plate III, f [upper spectrum]). The horizontal portion of the quartz tube was now strongly heated; the color of the discharge changed and the lower spectrum was obtained, with the band much weakened and the lines strongly developed.

The band spectrum has been photographed in the fifth-order spectrum of a seven-inch plane grating with a lens of three meters focus. A portion of the spectrum in the region of the 5626 line is reproduced on Plate III, g, in coincidence with the absorption spectrum. Since the photograph is reproduced as a negative, the absorption lines appear light instead of dark (upper spectrum). It is clear from the photograph that the spectra are not complementary, though every emission line of the band has an absorption line in coincidence with it. There are many absorption lines, however, which are not represented in the emission spectrum. This very probably results from something in the nature of dissociation. We have not yet studied the possible changes in the fine structure of the emission band spectrum with varying current-density, but there are indications that such changes occur, for some of our plates show greater dissimilarity between emission and absorption than the one just mentioned—h for example.

The action of high temperature on the band absorption spectrum has been investigated, however, and changes have been noticed which throw some light on the matter.

EFFECTS OF HIGH TEMPERATURES ON THE ABSORPTION SPECTRUM

Evans has given, in the *Astrophysical Journal* (*32*, 1, 1910) an account of an investigation of the disappearance of the absorption spectrum of iodine at high temperatures. His investigations were made with a spectroscope of low dispersion, and only the gradual disappearance of the bands was recorded. His results would be perfectly explained on the supposition that the absorption spectrum results from diatomic molecules I_2, which, at high temperatures, break down into monatomic molecules devoid of absorbing power. He found that the denser the vapor the higher the temperature necessary to cause the complete disappearance of the spectrum.

We have studied the phenomenon with a spectroscope of the Littrow type of six meters focus, using the fifth-order spectrum of the seven-inch grating. This grating, which is the best ruled by Anderson, gives very nearly its full theoretical resolving power (450,000). Photographs were made with dense iodine vapor in a quartz bulb (previously exhausted) heated by two blast lamps, and with the same bulb at room temperature, with much less dense vapor. We also made a large number of plates with a tube of pyrex glass heated in an electric oven. In every case we attempted to secure pairs of plates which showed the absorption spectrum at about the same degree of intensity, as this condition brought out the changes in the minute structure to better advantage. On the assumption that diatomic absorbing iodine breaks down into a colorless monatomic gas, we should expect the spectrum to fade away precisely as it does when the amount of vapor is decreased by lowering the density. This is not the case, however, as will be seen by comparing the photographs (positives) reproduced on Plate III, *i*, the lower one taken with the quartz bulb at high temperature (perhaps 1000° C.), the upper with the bulb at a temperature of 35°. The stem of the bulb was immersed in boiling water in the first case; consequently the iodine was at a density corresponding to 100°. If we compare the two photographs, we notice that some lines are much stronger in the upper spectrum of the cold vapor than in the lower. Some of these lines have been indicated by arrows. Many lines have about the same intensity in both spectra. Others, however, are distinctly stronger in the spectrum of the hot vapor. These also are indicated by arrows and small dots.

It appears, then, that the lines are affected in different degrees by an elevation of temperature. Those indicated by the arrows above the upper spectrum are the most readily quenched, and those indicated by arrows below the lower spectrum are the ones most resistant to temperature. Clearly we are dealing with something more complicated that the dissociation of a diatomic molecule. Similar differences were found in the case of the spectra made with the tubes of pyrex glass at temperatures ranging from 350° to 500° C.

It seems quite possible that the band emission spectrum will be found to be much more nearly the exact complement of the absorption spectrum of the vapor at a high temperature than in the case shown in Plate III, g and h. This matter will form the subject of a future investigation.

THE LINE SPECTRUM

Though the wave-lengths which we have redetermined are probably not much more accurate than those given by Konen, it appears to be worth while to give them, as we have divided the lines into two groups, the arc and spark lines previously alluded to. Moreover, we have determined the wave-lengths of some fifty of the lines from plates made in the fourth-order spectrum of the three-meter spectrograph, and these are correct to 0.01 A. It is doubtful if the others can be relied on beyond 0.1 A, and the same is true of Konen's values.

In Table I the 'arc' lines, or the ones which show a *decrease* of intensity as a result of increasing the current-density by a condenser or parallel spark-gap, are indicated by an asterisk. It should be noticed, however, that this effect is of variable magnitude, some lines being greatly weakened, others less so; some lines show no change at all, and others (the spark lines) are enhanced in varying degrees. On this account it is difficult to make a very sharp classification. All wave-lengths are given on the international scale and the values in italics were determined from plates made in the fourth-order spectrum and are correct to within about 0.01 A. The others are correct only to 0.1 A.

TABLE I

Wave-Lengths	Intensities		Wave-Lengths	Intensities	
	Without Spark	With Spark		Without Spark	With Spark
4632.4	4	10	4924.4	1	2
4634.8	1	8	4929.9	1	2
4640.7	2	10	4938.6	1	6
4657.4	1	6	4943.1	1	6
4663.8	1	6	4957.6	1	6
4666.5	4	12	4965.7	0	2
4675.5	6	10	*4968.33*	2	6
4676.5	1	8	*4974.5	1	0
*4687.3	1	0	4984.4	1	2
*4691.1	1	0	*4986.95*	2	10
4700.8	0	0	*4991.9	1	0
4702.5	1	2	4992.2	1	2
4707.9	1	2	5008.4	0	4
4711.7	1	2	5028.8	1	2
*4722.1	1	0	*5032.3	1	0
*4726.3	1	0	5036.1	1	6
4730.5	1	8	5046.4	1	4
*4734.1	1	0	*5048.1	1	0
4737.1	1	2	5057.4	1	4
4742.9	1	2	5061.9	0	4
4752.7	1	2	5065.5	2	6
4763.4	10	10	*5068.2	1	0
4765.7	1	4	5090.7	1	4
4768.2	0	6	5098.8	1	2
*4773.1	1	0	*5114.44*	1	8
*4775.8	1	0	**5119.32*	20	15
*4782.5	1	0	5124.6	1	2
4784.8	1	4	*5130.5	1	0
4787.2	1	2	5131.3	1	2
4788.2	1	2	*5133.2	1	0
4790.9	0	2	*5136.1	1	0
4799.8	0	4	*5138.5	1	0
4800.2	2	4	*5145.2	1	0
4806.4	2	6	5147.4	0	6
4808.0	0	2	5149.7	0	6
4828.3	2	6	5154.9	1	2
4835.1	2	6	5156.4	1	8
4850.4	2	10	*5161.20*	8	30
4853.1	1	2	5174.6	1	2

TABLE I—Continued

Wave-Lengths	Intensities Without Spark	Intensities With Spark	Wave-Lengths	Intensities Without Spark	Intensities With Spark
*4862.33	20	16	5175.1	1	2
4864.5	1	6	5176.3	1	2
4881.6	1	4	5178.1	1	8
4883.7	0	8	5185.14	1	8
*4887.7	2	0	*5186.3	1	0
4891.3	1	6	5189.4	1	2
4893.8	1	4	5198.9	1	8
*4896.72	12	8	*5204.08	10	4
*4902.2	4	0	5205.5	1	2
4908.5	1	2	5214.04	1	4
4910.3	1	2	5216.22	2	10
*4916.94	16	10	5228.93	0	8
*5234.58	10	8	*5501.00	2	0
5245.65	4	15	5504.77	2	8
5265.150 } 5265.266	2	10	5522.1	0	4
			5527.5	1	4
5266.8	2	2	5546.4	2	2
5269.36	2	10	5551.7	0	4
5288.7	0	4	5568.7	0	0
5296.7	1	2	*5586.3	4	2
5299.68	0	6	5590.3	2	2
*5304.3	1	0	5593.09	0	4
5309.0	1	8	5598.55 } 5598.68	2	6
5314.6	1	4			
5322.71	0	6	5600.21	2	6
5326.4	1	4	*5601.8	2	1
5336.6	1	0	5603.2	2	4
5338.20	6	18	5612.82	2	6
*5341.8	1	0	5625.66	4	15
5345.17	6	18	5643.4	1	4
5349.7	1	2	5673.7	1	4
5351.9	1	2	5678.06 } 5678.15	2	10
5356.0	1	4			
5367.5	2	4	5679.9	0	0
5369.75	4	12	5690.89 } 5690.96	2	10
5372.5	1	4			
*5374.5	1	0	5702.07	0	2
5380.1	1	4	5710.43	2	10

TABLE I—Continued

Wave-Lengths	Intensities		Wave-Lengths	Intensities	
	Without Spark	With Spark		Without Spark	With Spark
5405.11 ⎫ 5405.23 5405.38 5405.59 ⎭	4	16	5723.5 5725.0 5738.5 5739.5	0 0 2 0	0 0 10 10
5407.35	2	12	5734.8	0	1
5411.7	1	4	5760.8	2	8
5415.0	0	4	*5764.3	6	4
5421.97	0	4	5774.7	2	10
5422.71	0	4	*5780.4	2	1
*5427.4	6	4	5787.1	1	6
5435.80	4	10	*5790.2	1	0
5437.97	2	8	*5793.0	1	0
5449.0	1	4	5819.6	1	2
5457.1	2	4	5830.0	1	6
5464.77	6	20	*5832.7	1	0
5468.1	1	2	5875.1	1	4
5475.1		0	*5893.8	8	6
5479.55	1	6	*5908.5	1	0
5491.52	1	8	5920.7	1	4
5493.45 ⎫ 5493.05 ⎭	0	8	*5928.6	1	0
			5950.1	4	10
5497.08 5496.96 5496.85 5496.79 5496.73	2	15	*5956.6 *5960.0 5962.8 *5966.1 *5967.7	2 2 0 2 2	0 1 1 1 1
*5980.5	2	0	6257.4	1	4
*5984.2	2	0	6267.1	0	1
6007.6	1	2	6268.5	0	4
6015.8	1	4	*6276.8	1	0
*6023.9	6	2	*6280.3	1	0
*6036.5	1	0	6290.4	0	1
*6038.6	1	0	6291.3	1	2
*6041.4	1	0	*6293.9	6	2
6043.9	1	2	*6296.4	2	0
*6046.5	1	0	*6313.1	2	1
*6048.4	2	0	6320.9	1	4
*6053.0	2	0	6323.6	0	1
6068.8	1	4	*6330.2	2	0

TABLE I—*Continued*

Wave-Lengths	Intensities		Wave-Lengths	Intensities	
	Without Spark	With Spark		Without Spark	With Spark
6074.9	2	6	*6333.5	2	0
6078.2	1	2	*6337.9	4	2
*6082.3	10	6	*6339.5	6	2
6084.7	1	2	*6348.3	1	0
6086.8	1	2	*6350.9	1	0
6115.7	0	1	*6355.4	2	1
6125.4	1	2	*6359.1	4	2
6127.4	2	8	*6367.2	2	0
6132.9	1	2	*6371.6	2	0
6149.0	1	2	6375.8	0	1
6161.9	1	2	6378.2	0	1
*6187.0	1	0	*6411.1	2	1
*6191.6	4	2	*6415.2	2	1
6195.5	1	4	*6428.7	1	0
6200.4	1	4	6440.2	1	4
6204.7	0	6	6476.0	1	2
*6213.0	4	2	*6488.1	4	2
6229.2	0	2	6495.0	1	2
6232.9	1	2	6516.1	1	2
*6233.2	2	1	*6538.3	2	1
6236.3	1	4	*6560.3	4	2
*6240.2	2	0	6574.8	0	1
*6244.3	4	2	6578.0	0	4
6245.8	2	2	6579.8	0	1
6250.6	0	2	*6583.2	4	0
6255.5	1	2	6585.0	0	4

STUDY OF THE COMPLEX LINES WITH THE ECHELON

Instruments and methods. The echelon used in this work was a new one by Hilger, with twenty plates, 1 cm. thick, in optical contact. It was used in conjunction with a collimator and telescope, each of nine feet focus. It may be well to point out that this is about the right focal length to furnish the full resolving power recorded on a photographic plate. The telescopes of shorter focus usually employed, while excellent for visual observations, give poor results when used for photographic work. A wooden box surrounded the echelon and the objectives of the collimotor and telescope, and the temperature within this inclosure was kept constant to within 0.1° C. by a toluole thermostat. The slit of the collimator was removed and its place taken by the second slit of a Hilger constant-deviation spectroscope, used as a monochromator. To make settings with this instrument we simply swung the nine-foot collimator a little to one side, and viewed the slit of the monochromator with a lens of high power, opening it wide for the purpose of identifying the line, and then gradually closing it while keeping the desired line always within the aperture. This method is far simpler and more satisfactory than attempting to form an image of one slit on another by means of a lens, and we were able to study separately lines the distance between which was only seven Ångstrom units. Some of the lines showed such irregular structure with the echelon that we suspected the presence of neighboring lines that were not removed by the monochromator. This was found to be the case. In some cases, as we subsequently found, two complex lines were so close together that they were passed simultaneously through the slit of the monochromator. To overcome these difficulties we crossed the echelon with a plane grating of 15,000 lines to the inch, placing it with its glass plates horizontal, between the collimator and grating. The collimator lens was an ordinary telescope objective of one meter focus, and the spectrum was formed by a Cooke photographic objective of four inches aperture and the same focal length as the collimator. The slit was reduced in length to about 0.1 mm. by means of two strips of tin foil fastened on the inside with soft wax. The beveled edges were on the outside, which is the proper design for a slit, though many instrument-makers reverse matters and give us beveled edges on the inside, which sometimes cause spurious lines by reflection of oblique rays. If the slit of the spectroscope was opened wide, the echelon spectra of the complex lines could be seen in the broad images of the slit. The echelon was leveled and brought to the proper position by observing these spectra. The slit was then closed until it was reduced practically to a needle hole, and the echelon spectra contracted to vertical rows of minute dots or single dots, each row representing a complex line and each single dot a simple line. In this way it was possible to photograph with the echelon the entire iodine spectrum from violet to red on a single plate. We even succeeded in photographing the nitrogen band spectrum in this way. A photograph of a portion of the spectrum taken with the grating-echelon combination is shown in coincidence with a, the spectrum taken with the grating alone, on Plate III, b. A smaller portion of the spectrum more highly enlarged is shown by c', the spectrum formed by the grating alone lying between the spectra formed by the echelon-grating combination. We shall take up now the structure of the various lines studied thus far, designating in each case whether the observations were made with the echelon alone or were from the plates made with the echelon crossed with the grating.

LINE STRUCTURE

In the case of many of the complex lines we made accurate determinations of the wave-lengths of the principal lines correct to 0.01Å from photographs made in the fourth-order spectrum of the plane-grating spectrograph of three meters focus. These photographs also served as a check in interpreting the results obtained with the echelon. A portion of one of these is reproduced on Plate III, c, showing the complex line λ 5497. The constants of the echelon were as follows:

$\mu D = 1.57493$, $C - D = 0.00410$, $D - F = 0.00996$, $F - G = 0.00837$

From these values we calculated the constants in the Hartmann formula,

$$\mu = \mu_0 + \frac{C}{(\lambda - \lambda_0)}$$

$C = 141.09$, $\mu_0 = 1519$, $\mu_0 = 1.54267$.

The wave-length intervals corresponding to the distance of two successive orders were calculated from the formula

$$d\lambda_m = \lambda^2 \left\{ \frac{1}{t\left[(\mu - 1) - \lambda \frac{d\mu}{d\lambda}\right]} \right\}$$

for various wave-lengths, and their values are given in Table II. A curve showing the relation between λ and $d\lambda_m$ was then drawn, from which the wave-length intervals between successive orders for any value of λ could be found.

TABLE II

λ	$d\lambda_m$
4632.4	0.3345
4666.0	0.3397
5016.2	0.4006
5162.0	0.4237
5345.0	0.4672
5464.6	0.4854
5625.0	0.5175
5691.0	0.5308
5875.0	0.5695

While some of the complex iodine lines showed an irregular structure, or appeared as close doublets, the majority exhibited a series of lines of four or five members decreasing in intensity and separation toward the region of short wave-lengths, thus ||┃┃┃. In the majority of cases the width of the group was less than the distance between the orders of spectra of the echelon. Obviously we cannot, on a single photograph, get a true record of the relative intensities of the lines making up the series, for, if we put the echelon on position of 'single order' for the brightest line (first member of series), the last line will appear too faint in comparison, as it will be at or near the position of double order. We usually adjusted the echelon so as to show the last or faintest line in 'single order'; this made the first line relatively weak, but gave us a better record of the series for measurement. Two or three of the lines showed a series as wide as, or a little wider than, the distance between orders of echelon spectra. In this case the last member of the series falls upon or beyond the first member seen in the next order. In these cases, however, our photographs made with the grating in the fourth-order spectrum indicated the presence of the last member of the series, though it was not quite resolved, and by carefully comparing these photographs with those made by the echelon it was usually possible to determine the series. In the list which follows, the strongest line is designated by 0.000 and the distance of the other components from this zero position is indicated, the minus sign indicating of course the side of short wave-length.

Most of the lines in the violet prove to be single lines; the following showed structure:

λ4404. Four lines nearly equidistant, 0.000, −0.057, −0.105, −0.167 from plates made by echelon crossed with grating.

λ4465. Double line; 0.000, −0.069.

λ4474. Typical series of five lines of decreasing spacing and intensity. Strongest or main line 0.000, others at +0.078, +0.137, +0.190, and +0.232. This series appears to point toward longer wave-lengths. The other similar series point in the opposite direction. The reversal of this series was checked by a photograph made in the spectrum of the fifth order.

λ4632. Structure similar to foregoing, but series turned the other way. Main line 0.000, others −0.085, −0.152, −0.196, −0.228. This line studied by the echelon alone.

λ5065. Three components, +0.130, 0.000, −0.095 (echelon crossed with grating), middle component strong.

λ5161.20. Series of five lines. Main line determined in fourth-order spectrum of grating. Components at −0.105, −0.186, −0.241, and −0.276 (by echelon alone).

λ5245.65. Series of five lines. Main line determined by grating, which did not quite resolve the series. Echelon gave components at −0.053, −0.102, −0.146, and −0.192.

λ5265.150, 5265.266. Double line by grating. Separation of components by echelon 0.119.

λ5338. Close triplet by echelon alone. Main line strongest. Components at −0.041 and −0.083.

λ5345. Components 1 and 2 faint, 4 and 5 fairly strong. Calling No. 3 the main line, we have for the structure (5 seems to be double) +0.098, +0.029, 0.000, −0.051, −0.10, −0.116.

λ5356. Echelon crossed with grating shows five components spaced at nearly equal intervals, the width of the whole group being about 0.3 A.

λ5370. Series of five lines by echelon. Main line 5369.75 (by grating). Components of decreasing intensity at −0.054, −0.099, −0.146, and −0.192.

λ5405.59. Main line and series of decreasing intensity. Four of the members of the series resolved by grating, namely, .59, .38, .23, .11. The series was a little wider than the distance between the orders of the echelon, the last member (the fifth) falling beyond the main line in the next order. The series would be shown to better advantage by echelon plates of 7 mm. thickness. It could be studied only with the echelon crossed with the grating on account of the proximity of 5407 (a four-term series). It is represented thus: 0.00, −0.21, −0.36, −0.48, −0.55. (The distance between echelon orders in this region is 0.473.)

λ5407. Echelon crossed with grating shows series of four lines of decreasing intensity, 0.000, −0.062, −0.115, −0.160.

λ5436 is shown to be single, by echelon crossed with grating, and 5438 similar to the 5407 series but with closer spacing, 0.000, −0.05, −0.082, −0.115, and −0.161.

λ5464.77. Series of five components by echelon alone. Four shown with grating in fourth order. The first and strongest has wave-length given above, and the four others are of decreasing intensity and located at −0.106, −0.190, −0.225, −0.275.

λ5491.50 is single; 5493.45 and 5494.05 (by grating) appear with echelon crossed with grating as close doublet of about 0.13 separation—an example of a spurious result due to a confusing of orders.

λ5497.08. Main line of five member series. Grating measurements gave others as 5496.96, .85, .79, and .73. On account of proximity of other lines, it could be further studied only with echelon crossed with grating, which gave 0.000, −0.134, −0.233, −0.310, and −0.343.

λ5598.6. A triplet, by echelon crossed with grating. Components 0.000, −0.185, and −0.310 (the latter faint).

λ5600.2. Close doublet, separation of components about 0.05 Å.

λ5603. Three components by echelon crossed with grating: 0.000, −0.082, and −0.159.

λ5612.8. A doublet with components 0.068 Å apart.

λ5678. Doublet by echelon alone. Wave-lengths also determined by grating 5678.06 and 5678.15. Echelon gave 0.88 separation.

λ5691. Echelon alone gave a bright line with fainter companion 0.078 Å toward red, and a very faint one at −0.10 toward violet. The grating alone gave 5690.89 and 5690.96.

λ5710. A very complicated line. A strong triplet series with another strong line well separated from it on short wave-length side, and two or three fainter companions:

triplet	strong
+0.066, 0.000, −0.045, −0.073,	−0.110, −0.179

λ5738.5. Single.

λ5739.5. A triplet.

λ5774.7. Five components by echelon crossed with grating. The arrangement of the dots suggests that we may have two superposed lines, as the dots are not quite in line.

The structure of these lines are shown by Fig. 3. The reversal of the series in the case of the line 4474 is of especial interest, as it was the only case found. It is also worthy of mention that all of the complex lines belong to the 'spark' type. The width of

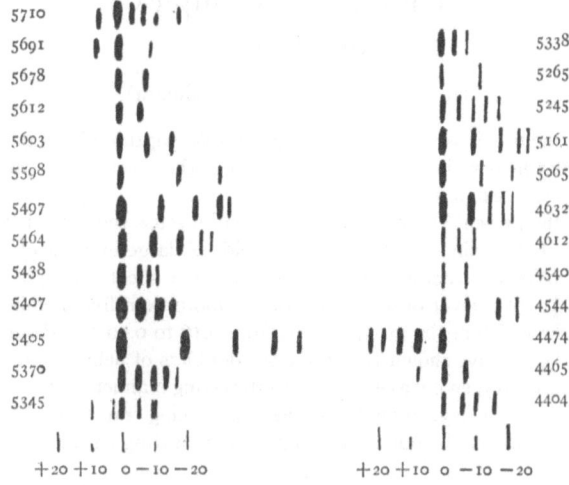

FIGURE 3

the quadruplet series 5438 is only a little more than one-half of the distance between the first two members of the quintuplet series 5405, in which the separation of the members reaches its maximum value. In the following paper the behavior of these complex lines in a magnetic field will be discussed.

No. 6

Zeeman-Effect for Complex Lines of Iodine

(In collaboration with M. Kimura)

In the previous paper we have given a description of the many complex lines which we have discovered in the spectrum of iodine electrically excited in vacuum tubes.

In the present communication we shall discuss the behavior of some of these lines in the magnetic field. A large number of the lines have a structure resembling that of a short series, thus, ‖│▌▌. We never observed a series of more than five members, and the width of the group varied from 0.16 to 0.50 Å. Paschen and Back [1] have shown that the close doublets of helium and the triplets of oxygen behave in a most interesting manner in the magnetic field, the oxygen triplet, for example, being transformed into a single line for the polarized component vibrating parallel to the field.

Our preliminary observations, which were made without polarizing apparatus, showed that the lines which had the structure figured above gave, in strong fields, a triplet of quite normal appearance. In weak fields the structure became too complicated to follow, and we immediately resorted to a polarizing apparatus by which the components vibrating parallel and perpendicular to the field could be studied separately. The tubes employed were of the same type as that described in the previous paper. The short capillary part of the vacuum tube was mounted between flat pole pieces, the field being essentially homogeneous in the region of the capillary. The tube was observed 'end-on', i. e., in a direction perpendicular to the lines of force. A natural crystal of Iceland spar about 1.5 cm. thick was used as a double-image polarizing prism. This was placed close to the capillary, and, when properly oriented, gave two polarized images very close

[1] *Annalen der Physik*, 39, 897, 1912.

together, the one immediately above the other. Real images of these were formed on the first slit of the Hilger constant-deviation spectroscope by means of a photographic objective of high quality. These images were about a millimeter in diameter, as we placed the lens much nearer the tube than the slit, and were separated by a distance of about 3 mm. Both polarized components could thus be photographed with the echelon simultaneously, and by raising the images about 1.5 mm. on the slit we could obtain two more records showing the unmagnetized lines in coincidence with the magnetized ones. Various methods of bringing about this vertical shift were tried, but all were found unsatisfactory until the following simple expedient was adopted. A piece of plane-parallel glass (Michelson interferometer flat) was mounted a short distance from the slit of the Hilger spectroscope and arranged to rotate on a horizontal axis, by which it was possible to incline the plate from the vertical at the angle necessary to produce the requisite shift. A graduated paper scale and a light lever attached with sealing-wax, to give the desired rotation, completed the apparatus. The advantage of the plate is that it shifts the two converging beams *without changing their direction*. Acute prisms, and other devices which we tried, changed the direction of the beams, reducing or destroying entirely the illumination of the echelon. In taking our plates we recorded the current flowing in the magnet in each case, and subsequently determined our fields by comparing two deflections of a ballistic galvanometer, produced respectively by the quick removal of a small exploring coil from between the pole pieces and from the center of a standard solenoid giving a known field. This method could not be used for the stronger fields (above 1000 gauss), as the field in the coil was only 860 gauss with a current of 18 amperes. The stronger fields were measured by observing the Zeeman-effect on the green helium line 5016 (which is known to exhibit the normal effect) from the formula

$$H = \frac{\Delta\lambda}{0.94\,\lambda^2}\,10^4,$$

in which the wave-length is expressed in centimeters, and $\Delta\lambda$ is the separation of the outer components of the Zeeman triplet in a field of H gauss.

The following lines were studied: λλ 5464, 5161, 4632, these having five components each, in the form of a series of decreasing intensity and spacing; λλ 5338 and 5345, each a close triplet; 5691, a doublet having one strong and one weak component; and 5624, a single line.

Photographs of many of the other lines were made, but only those mentioned above were measured. The resolving power of the echelon was not quite sufficient (twenty plates of 10 mm. each) and as a future investigation with a forty-plate echelon is contemplated by one of us, the results given in the present paper are to be regarded as preliminary in their nature.

λ 5691

We shall begin with a discussion of the action of the magnetic field on this line, as its structure is very simple, a strong line with a fainter companion 0.078 A on the side of longer wave-lengths. The behavior of the perpendicular and parallel components of polarization in fields of increasing strength is shown on Plate IV, *j*. Shorter wave-lengths are to the right in these figures.

In the case of the parallel component the main line remained undisplaced, as in the normal Zeeman-effect up to a field of about 2000 gauss, but its intensity increased and it showed a distinct broadening. The satellite, however, approached it and increased in intensity until, at about 2500 gauss, the intensities appeared about equal and the lines almost fused. At 3220 they had completely united into a single line with a wave-length intermediate between that of the main line and the satellite, with a faint companion on the side of short wave-lengths; with increasing field the bright line suffered a further displacement toward the red, the faint companion moving toward the violet and disappearing in fields above 5000. Over a dozen plates in all were made and measured, and the results are shown in the form of a graph in Fig. 1.

In the case of the perpendicular component, the main line was doubled, the separation being normal up to a field of about 2000 gauss, the satellite remaining unaffected. With an increase of field the satellite was deflected as if pushed along by the positive branch of the doublet, the two finally fusing as shown in Fig. 1. As will be seen from the diagram, the resulting doublet continues to widen and at 9000 gauss is symmetrical with respect to a point midway between the main line and the satellite. This phenomenon of the fusing of the satellite with the main line in the case of the ⊥ component, and with one of the branches of the doublet in the case of the ⊥ component, was observed also in the case of other iodine lines, and is in agreement with observations made by Nagaoka and Takamine on the lines of other elements. In strong fields, and in the absence of the polarizing apparatus, we have a triplet, with its central component displaced (from the position originally occupied by the main line for zero field) toward the red.

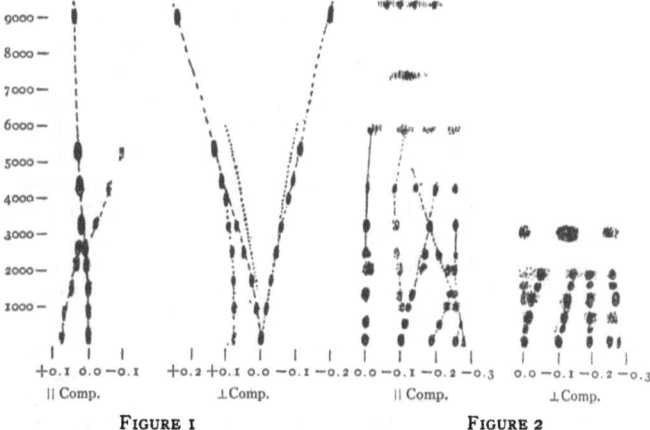

FIGURE 1   FIGURE 2

The dotted lines in Fig. 1 (⊥ component) indicate the calculated separation for the normal Zeeman-effect.

λ 5464.77

This line consists of five components, decreasing in intensity and separation toward the side of short wave-length, suggesting a miniature Balmer series. The wave-length of the main line was determined by the grating, and the components are located at −0.106, −0.190, −0.255, and −0.275, the latter being very faint and not appearing resolved in the reproduction. The appearance of the composite line is shown on Plate IV, k. For each field-strength the parallel and perpendicular components of polarization are shown in coincidence with the line in zero field.

We will consider first the behavior of the components of vibration parallel to the field. Notwithstanding the fact that the parallel component is unaffected by the field in the normal Zeeman-effect, in the case of this complex line, the three members of shorter wave-length are so sensitive that even the residual magnetism of the electromagnet, after the current was shut off, was sufficient to alter the appearance of the series in a very marked manner, and this for the *parallel* component, ordinarily uninfluenced. On this account it was necessary to demagnetize the magnet very carefully, by repeatedly reversing and diminishing the current, until no appreciable attractive force was exerted by the poles for a piece of soft iron. It was also important not to pass from a strong to a weak field without demagnetization.

The behavior of the lines, when only the parallel components of vibration are recorded, is shown by Fig. 2. In a field of only 150 gauss a very distinct effect was observed on the two lines of shortest wave-length (-0.255 and -0.275), and at 500 gauss they fused into a single line. The line at -0.190 first widens, and at 600 gauss becomes double. We now have five lines, as in the beginning, though with a different spacing and distribution of intensity. If this were an isolated observation, one might erroneously conclude that the magnetic field had merely pushed the lines closer together. As the field increased in strength the negative branch approached, and finally fused with, the line formed by the fusion of the two referred to above, which appears to move slightly in the positive direction to meet the other line. The other branch (positive) formed by the division of -0.190 could not be followed beyond 600 gauss. The further behavior of these lines with increasing field is well indicated by the figure. The line formed by the fusion of the three lines just referred to divides again, one member increasing its wave-length, the other remaining fixed. The component at -0.106 divides as indicated, the negative branch, which is strongly displaced, fusing with the positive branch mentioned above. The positive branch, which is much fainter, remains almost in coincidence with the original line. These changes can be followed on Plate IV, *k* (upper figures marked ||). Above 3000 gauss it is difficult to interpret the plates, as the components become hazy. At 3400 gauss we have four lines, and at 4400 gauss five lines again, the probable manner of transition being indicated in Fig. 2. At 6000 gauss we again have but four lines (hazy). At 7500 we found a continuous background, with a hazy line in the center, but we were unable to trace out the transition.

The perpendicular component of polarization is affected in a very different manner. The third line of the series remains undisplaced in fields below 2000 gauss, the first and second lines giving strongly displaced negative branches and very faint positive branches, which show little displacement and are difficult to follow. At 3200 there is a strong central component and two lateral fainter components, but we have not been able to determine from the plates just how the transition takes place, nor have we followed the development of the hazy doublet which appears in very strong fields.

$\lambda\,4632$

The behavior of this line was studied in fields up to 4000 gauss and is shown graphically in Fig. 3. Normally it is a five-line series similar to $\lambda\,5464$, but it is affected in a different manner. For the parallel component line No. 1 broadens and shows a suggestion of doubling. No. 2 doubles, the positive branch disappearing in fields above 2000, the negative remaining up to over 3000. No. 3 is unaffected up to 1000 gauss, then deviates rapidly in a negative direction. No. 4 is also displaced in the same direction. At 3900 gauss we have a doublet. For the perpendicular component of polarization, lines Nos. 1 and 3 double and 2 and 5 remain unaffected. The resolving power of the echelon was insufficient to accomplish more than a suggestion of this doubling, however, in the widened lines. At 4900 we have a triplet with a strong middle component. The behavior of the line is shown on Plate IV, *l*.

$\lambda\,5161$

The parallel component only was studied in the case of this line, which is a five-line series similar to $\lambda\,4632$ and $\lambda\,5464$. Its behavior is shown by Fig. 4. The first line of the series exhibited only a slight widening and slight displacement toward the violet as the field-strength increased. The second line behaved in a remarkable manner. It became distinctly double in a field of 1000 gauss, and at 1820 the two components were widely separated. The doubling was not symmetrical, however, for the positive branch attained its maximum displacement at 1820, while the negative branch continued to move toward the violet with increasing field. The positive branch eventually fuses with line No. 1, which moves over to meet it (4000 gauss). The displacement of the negative branch was proportional to the field-strength up to 3500 gauss. Line No. 3 was very slightly displaced toward the violet with increasing field and fused with the negative

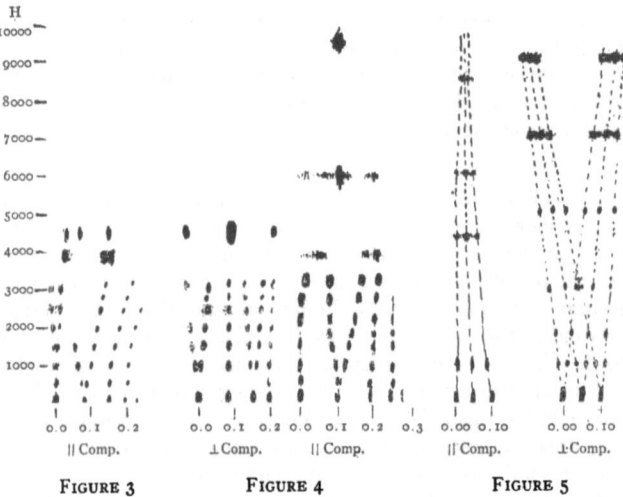

FIGURE 3   FIGURE 4   FIGURE 5

branch of No. 2 at 5000 gauss. Lines No. 4 and 5 fused at 500 gauss and faded away above 3000 gauss. At 6000 gauss we have a broad hazy line in the position of line No. 2, and at 9700 gauss it is found slightly displaced from this position toward the violet. These changes are shown on Plate IV, *m* and *n*, the latter figure showing the group in ten different stages, from zero field up to 9700 gauss, with the photographs mounted in coincidence.

$$\lambda\ 5338$$

In a weak field the main line of this triplet was decomposed into a triplet with normal separation and polarization. The satellite at −0.041 was also decomposed into a triplet, but the component lying on the side toward the main line suffered a greater displacement than the one lying on the other side of the central component. The same was true of the satellite at −0.083 except that the dissymmetry was even greater.

In strong fields we have a diffuse triplet which forms in the manner indicated by Fig. 5, which is, however, the graph for the similar line λ 5345.

λ 5345

The behavior of this line is similar to that of λ 5338, and the measurements made from the plates are recorded in Fig. 5.

λ 5625

This is a single line and gave a symmetrical triplet, with a separation somewhat greater than that of a normal triplet.

For normal triplet $\frac{\nabla\lambda}{\lambda^2 H} = 0.94 \times 10^{-4}$.

For λ 5625, $= 1.26 \times 10^{-4}$,

that is, in the ratio 3 : 4.

The chief points of interest which have been brought out in this investigation may be summed up as follows:

The complex lines having the form of a series with decreasing intensity and separation are not at all affected in a similar manner by the magnetic field.

In the case of any given complex line the components are affected to very different degrees. Certain components may not be affected at all, while others break up into doublets, the components of which sometimes fuse with neighboring components and sometimes fade gradually away as the field-strength increases.

In the case of the perpendicular components of polarization we have not traced the development of the widely separated hazy doublet which appears with strong fields from the complex which develops in weak fields. This will require a somewhat higher resolving power than that available in the present work. Obviously the method of the non-homogeneous field would be especially adapted to the study of these complex lines, as the transition could then be traced by very gradual steps. We made some experiments along these lines, with flattened capillary tubes and pointed poles, but the results were not very satisfactory. With the tubes used in the latter part of the work, with internal electrodes, described in the previous paper, it seems probable that excellent results can be obtained in this way.

No. 7

A Photometric Study of the Fluorescence of Iodine Vapor

(In collaboration with W. P. Speas)

The reduction in the intensity of the fluorescence of iodine vapor caused by the admixture of air or other foreign gas was studied by one of the present writers a number of years ago (Wood, *Phil. Mag.* xxi. p. 309, 1911); and subsequently Wood and Franck (*Phil. Mag.* xxi. p. 314, 1911) discovered that the gases which were strongly electro-negative were the most effective in reducing the intensity of the fluorescence. Of all the gases studied the least effective was helium, the intensity of the fluorescence of the iodine vapor, when mixed with helium, even at two or three centimeters' pressure, being almost as great as *in vacuo*. The color of the fluorescent light was changed, however, from yellowish-green to orange-red by the presence of the helium; and the curves obtained showed that this resulted from the circumstance that the helium reduced the intensity of the radiations of shorter wave-length in the fluorescent spectrum to a greater degree than the less refrangible radiations. The extensive investigations of the remarkable resonance spectra emitted by the vapor when excited by monochromatic light, which have been carried on during the past two years by one of us, made a further photometric study desirable; for it appeared probable that careful determinations of the variation of the intensity of the radiations with the density of the iodine vapor would throw some light upon certain obscure points: for example, the circumstance that the faint band-spectrum which accompanies the resonance spectrum is more strongly developed when the iodine vapor is at very low density, the tube being cooled by ice. Moreover, it is of considerable interest to determine to what extent the luminosity of an iodine molecule is diminished by the proximity of other iodine molecules; in other words, to determine the effect of *iodine* vapor at different pressures upon the intensity of the iodine fluorescence, for comparison with the effects of the various other gases determined in the earlier work.

In the present case, however, the matter is complicated by the circumstance that an increase of pressure increases the number of fluorescing molecules.

It has been found possible, however, to allow for this circumstance and construct a curve showing the destructive action of iodine vapor upon the fluorescence of iodine vapor, precisely analogous to the curves constructed for helium, argon, nitrogen, hydrogen, etc., in the earlier investigation.

The iodine vapor was contained in an exhausted glass tube of the same form as those used in the study of the resonance spectra. The image of a quartz mercury arc formed along the axis of the tube by a large condenser excited a fluorescence of very constant intensity, which was measured by a photometer viewing the fluorescent vapor column 'end-on'.

The photometer was of the same type as that used in the earlier work, the comparison source being a white screen illuminated by the light of a Welsbach mantle passed through suitable filters for the purpose of matching the yellowish-green color of the fluorescence. The temperature of the tube was raised by a water-bath, or lowered by the immersion of a small lateral tube in a bath of alcohol contained in a small Dewar cup and cooled to any desired temperature by the addition of liquid air. The density of the iodine vapor is determined by the temperature of the coldest part of the system, so that when working below room-temperature it was necessary to vary only the temperature of the small lateral tube.

It was found that a measurable fluorescence was obtained even with a density corresponding to $-30°$ C. It was impossible, however, to obtain an absolutely black background, even with the end of the tube painted black for a distance of 10 cm. To determine the small amount of diffused and reflected light sent out by the background, it was only necessary to immerse the lateral tube in liquid air, which removed every trace of iodine vapor from the observation tube, and measure the intensity of the very feeble illumination of the background. This constant quantity was subtracted in each case from the measured intensity of the fluorescence.

When working above room-temperature, the entire tube was immersed in a rectangular glass tank filled with distilled water at the desired temperature.

The variation of the intensity of the fluorescence as a function of temperature is shown by curve A, fig. 1, ordinates representing

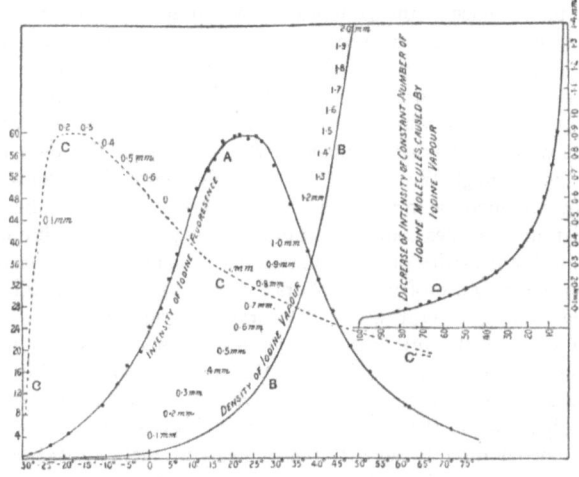

FIGURE 1

intensities and abscissæ temperatures. It is to be understood, of course, that the alteration of the temperature influences the intensity of the fluorescence only by changing the vapor-density.

The variation of the density of the iodine vapor with the temperature is shown by curve B, the pressures in millimeters of mercury (ordinates) being indicated to the left of the curve. This curve was plotted from the values obtained by Baxter, Hickey, and Holmes (J. American Chemical Soc. xxix. p. 127, Feb. 1907) between 0° C. and 55° C. They give no values below 0°, though a faint fluorescence can be observed at —30°, to which point I have carried the curve by rough exterpolation.

The intensity curve A is remarkably symmetrical, with its maximum at 20°–25°. As we increase the temperature from — 30° to 0° the intensity increases, probably very nearly in proportion to the increase in pressure, since at these very small densities the intensity of the radiation given out by a molecule is not diminished by the presence of its neighbors. Above 0°, however, the increase is no longer proportional to the increase in the number of molecules, since the vapor begins to destroy its own fluorescence in the same way as would another gas, such as nitrogen or carbonic acid, only to a very much greater degree. For example, at a pressure of .05 mm. (at 5° C.) the intensity is 33, while at a pressure of 0.1 mm. (at 11° C.) the intensity is but 46, instead of 66, which is the value which we should expect if there were no interaction between the molecules. Between 17° and 25°, though the number of molecules more than doubles, there is no increase in the intensity, the increase in the number of radiating molecules being almost exactly compensated by diminution in the intensity of the radiation from each one which results from the presence of its neighbors. Above 25° the reduction of intensity preponderates, and the curve falls rapidly.

If we plot the intensities (ordinates) against pressures (abscissæ) we obtain curve C (dotted), the pressures (abscissæ) being recorded along the curve. This curve gives us a better idea of the phenomenon than curve A, since in this case the intensities are plotted directly against the changes in the physical state which influences the radiation. This curve shows us, however, only the change in the intensity of the *total radiation* emitted by all the molecules with increase of pressure.

In view of the previous work, in which the diminution in the intensity of the radiation from a constant number of iodine molecules resulting from the presence of foreign molecules was investigated, it is of great interest *to determine to what extent the radiation from a given group of iodine molecules is diminished by interpolating other molecules of the same kind*, or, in other words, the effect of iodine vapor in reducing its own fluorescence as compared with the effect of other gases.

This can be done very easily by combining the values shown by curves A and B in the following way.

At a temperature of 0° the pressure is .03 mm. and the intensity of the fluorescence is 24. We wish now to determine the intensity of the radiation of this *same group of molecules*, when an equal number of similar molecules has been interpolated. We raise the temperature to 7°, the pressure doubles (.06 mm.), and the intensity increases to 36. We have, however, measured the radiation from all of the molecules, and we are concerned only with that which is emitted by the original group, which contributes one-half of the measured intensity; consequently we must divide the 36 by 2, which gives us 18. The intensity of the radiation of the group has been reduced from 24 to 18 by an *increment* of pressure equal to .03 mm.

At a temperature of 11° the pressure is .09 mm. and the intensity is 47, one-third of which, or 15.7, is contributed by the original group; consequently the intensity is reduced from 24 to 15.7 by a pressure increment of .06 mm.

We can in this way construct a curve showing the decrease in intensity resulting from the interaction between the molecules.

It is to be noted, however, that we must choose, for the original group, a mass of vapor at a pressure below that at which the action of one molecule upon the radiation from a neighboring one is appreciable.

If we take as our starting-point the intensity 12, at a pressure of .015 mm. we find that at .03 mm. the intensity is 24: one-half of this is 12, our original value—in other words, *no reduction in intensity has resulted* from an increment of pressure of .015 mm. At .045 mm. the intensity is 30, one third of which is 10, a slight reduction having occurred.

In this way curve D was computed, the values calculated being multiplied by 8⅓ so as to make the intensity of the radiation from the vapor at the lowest pressure equal to 100. This curve shows us the extraordinary effect of iodine vapor upon its own fluorescence, the vapor at 1 mm. pressure reducing the intensity from 100 to 5.

It is interesting to compare the action of iodine vapor with that of the gases and vapors studied in the earlier work. In these other cases a constant temperature was employed, with the result that the iodine vapor density remained unchanged, and the intensity of the fluorescence was measured when various gases at different pressures were introduced into the bulb.

The intensity is reduced from 100 to 19 by hydrogen at 24 mm., by air at 11 mm., by CO_2 at 7 mm., by ether vapor at 3 mm., by chloride of iodine at 1.8 mm., and *by iodine vapor* at a pressure of only 0.4 mm.

It is probable that chlorine would be still more effective than iodine, as it is more strongly electro-negative. The value given above for chloride of iodine is the value given in the earlier paper for chlorine, the fact having been overlooked that the vapors unite to form the compound when mixed. There was always an excess of iodine in the bulb, so that there is little doubt but that the correct interpretation of the experiment is to consider the active vapor iodine chloride instead of iodine.

Bromine vapor is more electro-negative than iodine, and though it has an absorption spectrum similar to that of iodine, its action in destroying its own fluorescence is so powerful that it is only possible to observe fluorescence at pressures probably in the vicinity of .001 mm., the intensity then being so small that it is only with difficulty that the phenomenon can be detected. Sunlight must be focused at the centre of the exhausted bulb, and the bromine vapor condensed by applying solid CO_2 to the exterior: just before the last trace of vapor is condensed, there is a very feeble green fluorescence, of about the intensity of that shown by iodine vapor at $-30°$.

The results obtained with iodine vapor at varying pressure emphasizes the following general statement made in the earlier paper:

"In order to obtain a visible fluorescence we must have a sufficient number of molecules present: their number must not, however, be so great as to cause them to disturb each other. The pressure at which maximum fluorescence occurs depends upon the electrical character of the molecule."

ABSORPTION OF THE FLUORESCENT LIGHT BY IODINE VAPOR

It is obvious that, for a correct interpretation of the results found with the photometer, determinations of the absorbing power of the vapor for the fluorescent light must be made, since in all of the experiments the fluorescent light is obliged to traverse a greater or less amount of absorbing vapor.

The color of the fluorescent light is distinctly red with dense vapor, orange-yellow at room-temperature, and yellow with a suggestion of green at the lowest temperatures. While this change is due in part to absorption of the green portion of the spectrum of the emitted light, there is undoubtedly another factor at work. In the earlier investigation it was found that the color was changed very markedly to red by the admixture of helium with a constant amount of iodine vapor, the same effect being observed in decreasing degrees with argon, hydrogen, and nitrogen. No change of color was, however, observed when the intensity was reduced by chlorine. The suggestion was made that a foreign gas reduced the intensity of the fluorescence in two ways—by its electro-negative quality (the reduction in this case being unaccompanied by change of color), and by collisions, which reduced the intensity of the short waves more than that of the long. It is quite possible that the collisions weaken what have been termed the resonance radiation lines more than the lines of increased wave-length.

In the case of the weakening of intensity by iodine vapor, the change of color is probably largely due to absorption, since measurements showed that the fluorescent light was more strongly absorbed by iodine vapor than light of the same color obtained by filtering the light of the Welsbach light through suitable color-filters. This results from the circumstance that the fluorescent spectrum is discontinuous, some of its lines coinciding with absorption-lines. If the two fields of the photometer were matched, one being illuminated with the fluorescent light, the other with the filtered white light, the balance was destroyed if a bulb containing iodine vapor was held between the eye and the photometer. Measurements were also made by restricting the length of the illuminated column of iodine vapor by means of screens, illuminating first the end of the tube away from the photometer, and then the nearer end.

The actual intensity of the fluorescence was the same in the two cases, but in the former, owing to the greater thickness of the layer of iodine vapor traversed by the emitted light, the measured intensity was less.

The results indicated that the portions of the illuminated column nearest the photometer contributed more to the intensity than the portions farther away.

It was found that the absorption was much stronger for the fluorescence of the vapor at 0° than at room-temperature, amounting to 43 per cent. in the former case and 29 per cent. in the latter, for a layer of iodine vapor at 23° and 14 cm. in thickness. The absorption was brought about by inserting an exhausted bulb, 14 cm. in diameter, containing iodine crystals, between the fluorescent tube and the photometer. Instead of removing the bulb, to determine the intensity without absorption, the iodine vapor was condensed by the application of cotton wet with liquid air. In this way the loss due to reflexion by the walls of the bulb was eliminated.

The effect of absorption will be to cause a decrease in the intensity of the fluorescence with increasing vapor-density. The effect is somewhat complicated by the circumstance that the green portion of the spectrum is more strongly absorbed than the red. This will cause a change in the color of the fluorescence, apart from the cause already mentioned, namely collisions with other molecules. The intensity curve consequently falls more rapidly than it would if absorption were absent. It is not very easy to correct for absorption, since the light from each element of the column of vapor illuminated is obliged to traverse a different thickness of vapor.

The chief cause of the diminution of intensity is the mutual action between the molecules. In the case of denser iodine vapor there is no trace whatever of superficial fluorescence, or a glowing of a thin layer of the vapor in contact with the wall. This would be practically uninfluenced by absorption. It is present in a very marked degree with mercury vapor, both for the visible fluorescence, obtained by illuminating the dense vapor contained in a heated quartz bulb with the light of the spark, and the ultraviolet resonance radiation, stimulated at pressures below 1 mm. by the 2536 line of the mercury arc, as will be shown in one of the following papers.

No. 8

The Magneto-Optics of Iodine Vapors

(In collaboration with G. Ribaud)

The magnetic rotatory polarization of iodine vapor was discovered in 1906 by one of the present writers.[1]

A small glass bulb, highly exhausted and containing a small crystal of iodine, was placed between the perforated pole-pieces of a powerful electromagnet, and warmed until the iodine vapor showed a light purple tint. Polarized white light was now passed through the hollow cores of the magnet and the bulb, and received by a Nicol prism set for the position of extinction. On exciting the magnet the nicol transmitted light of a bright emerald-green color, as a result of the selective rotation of the vapor due to the presence of innumerable absorption lines in the yellow, green, and blue region of the spectrum. By means of a concave grating of 14 feet radius, the spectrum of the transmitted light was resolved into bright lines, equalling in narrowness the emission lines of the iron arc. Similar results were obtained with the vapor of sodium[2] and bromine,[3] and the spectra obtained in this way were named magnetic rotation spectra.

In the case of the vapor of sodium it was found that in the red and orange region, some of the absorption lines rotated the plane of polarization to the right, others to the left. This phenomenon was observed by employing a double prism of Fresnel (right and left handed quartz), which, when employed in the well-known manner of Macaluso and Corbino, causes the appearance of horizontal dark bands in the spectrum.

Selective rotation of the medium manifests itself by the penetration of light from the bright into the dark bands. In the red and orange region, bright needles of light were observed to shoot into the dark region as soon as the magnet was excited,

---

[1] R. W. Wood, *Philosophical Magazine*, xii, p. 329 (1906).
[2] R. W. Wood, *Philosophical Magazine*, xii, p. 499 (1906).
[3] G. Ribaud, *C. R.*, clv., p. 900 (1912).

some of them projecting themselves downward, others upward, indicating positive and negative rotations of the plane of polarization.

In the green and blue regions of the spectrum, the rotations, while sufficient to give a brilliant bright line spectrum, were insufficient to make observations with the double prism possible. It must be remembered that a rotation of 90° is necessary to cause the light to pass from the center of a bright band to the center of a dark one. In the case of both sodium and iodine comparatively few of the absorption lines, of which there are many thousand, appeared to rotate the plane of polarization to an appreciable degree, the magnetic rotation spectrum being made up of something over 100 lines altogether. In the case of bromine, observations made with a concave grating of 1.60 m. focal length (when used with a collimating lens) have shown bright lines (rotation lines) for all of the absorption lines which the grating was capable of resolving, in other words, the absorption spectrum and the magnetic rotation spectrum were complementary. This was, however, true only when the vapor was at very low density. At higher densities the appearances were totally different.

Recent work on the resonance spectra of iodine has shown that a resolving power of at least 300,000 is necessary for an exact study of the phenomena produced by all of the rays of absorption, which are extremely fine and very close together, over one hundred having been counted in region 6 A. U. in width (distance between the D lines), on a photograph made with the 42-foot plane grating spectrograph at East Hampton, N. Y.

The remarkable resonance spectra excited when the vapor is stimulated by monochromatic light of a frequency corresponding to that of one of these very fine absorption lines, have shown the importance of a study of the vapor in a magnetic field with a resolving power sufficient to clearly separate all of the lines.

The aim of the present investigation has been to determine the exact nature of the rotation produced by the lines of absorption, since the more recent investigations, just alluded to, have shown that the earlier results dealt with rotations produced by close groups of lines, no record having been obtained of the nature of the rotation to the right and left of a single line.

It was of especial importance to determine whether the rotation to the right and left of an absorption line was of the same nature, *i.e.*, either positive negative, as is the case with the *D* lines of sodium, or whether any case of anomalous rotation occurred, *i.e.*, positive on one side of the line and negative on the other.

Small glass bulbs about 2.5 cm. in diameter, highly exhausted and containing a crystal of iodine, were mounted between the poles of a large Weiss electromagnet. The bulbs were supported in a brass tube of 3 cm. internal diameter, furnished with two lateral holes for the passage of the light, and heated electrically by a spiral of nickel wire placed below the bulb. A cover of mica forced the heated air rising around the bulb to escape through the side holes, and prevented iodine crystals from depositing on the walls in the path of the beam of light, as was invariably the case if the brass tube was open at the top. The source of light was a quartz mercury arc arranged 'end on', the observations being restricted to the seven or more absorption lines of iodine which are covered by the broadened green mercury line.

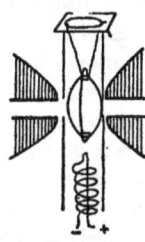

FIGURE 1

Our first observations were made with a six-inch plane grating in the fourth order spectrum (res. power 300,000) combined with a plano-convex lens of 4.20 m. focus. This lens was kindly placed at our disposal by M. Deslandres, director of the Meudon Observatory.

It was hoped that observations could be made with the Fresnel double prism, but preliminary experiments showed that the rotations were too small to cause any appreciable penetration of light into the region of the dark bands. It was necessary, therefore, to make use of the method previously used in the study of the magnetic rotation of sodium in the green region of the spectrum. The polarizing nicol is rotated until sufficient light is restored to render the absorption lines visible (10 to 15 degrees). The magnet is then excited and the spectrum brightens at the points where the rotation is in the opposite direction to that in which the nicol has been rotated, and darkens where the rotation is in the same direction.

Suppose that the rotation is positive to the right and left of one absorption line, and negative to the right and left of another. The former will appear narrower when the field is excited as a result of the brightening of the edges of the dark line. The latter will, however, appear broader than in the absence of the magnetic field. If the rotation is anomalous, the center of the absorption line will appear slightly shifted as a result of its becoming brighter on one side and darker on the other. This method of observation gives results quite as conclusive as those obtained with the Fresnel prisms, and is well adapted to cases where the maximum rotation is less than forty or fifty degrees.

It was found necessary to orient the analyzing nicol so as to obtain the maximum reflection from the grating, for the polarizing power of a grating is very large, especially in spectra of higher orders than the second. In the present instance the difference in the reflecting power was certainly five or six fold for vibrations parallel and perpendicular to the grooves. With this arrangement of the apparatus we had no difficulty in observing rotations both positive and negative in direction, but the intensity of light was not quite sufficient to enable us to be sure of what happened on both sides of the lines. The rotation appeared to be more marked on one side of the lines than on the other, and we felt uncertain about our results.

We accordingly substituted for the grating a very fine echelon, loaned through the kindness of Mr. F. Twyman, of the Hilger Company. This instrument consisted of 20 plates in optical contact (each plate 15 mm. thick), and gave a resolution equal or superior to that of the grating and an image of much greater intensity.

It was immediately obvious, with this instrument, that certain lines became broader when the magnet was excited, and that a reversal of the direction of the field caused them to become so fine as to be almost invisible, as a result of the brightening of the regions bordering them. This brightening was in some cases much stronger on one side of the line than on the other.

To obviate the necessity of reversing the field to observe these changes, a half-wave plate of mica was placed over one-half of the slit of the collimator, the principal directions of the plate coinciding with those of the polarizing prism. As the echelon showed a trace of astigmatism, it was necessary to form an image of the edge of the half-wave plate at a distance of a few millimeters behind the slit, in order to have a sharp hair-line divide the two fields of view.

The action of the half-wave plate is as follows:—

Let OP (Figure 2) represent the direction of vibration of the light traversing the iodine bulb. The principal directions of the mica plate oy and ox being parallel and perpendicular respectively to OP, the light traverses the plate without change. The analyzing nicol is now turned through an angle a from the position of extinction, and it transmits the component of OP parallel to OA.

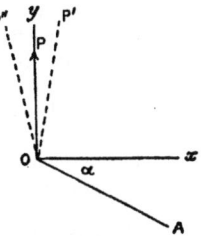

FIGURE 2

Suppose now that a certain wavelength λ is rotated by the magnetized iodine vapor from the position OP to OP', *i.e.*, in the same direction as that in which the analyzer was turned. In that part of the field of vision given by the light which has not traversed the mica, the wave-length λ will appear darker than before the excitation of the magnet. The vibration which traverses the mica (wave length λ) is rotated by the mica from the position OP' to OP'', and is consequently more copiously transmitted by the analyzer than the wave-lengths not rotated by the iodine. Consequently λ appears brighter in this part of the field of vision. The two conditions seen with magnetic fields of opposite direction are thus visible simultaneously one above the other, and in exact coincidence. Any shift, due to anomalous rotation, would thus be doubled, but no such shift was observed. The arrangement of the entire apparatus is shown in Fig. 3. It was at once evident that the observation of a larger rotation on one side of certain absorption lines than on the other, made with the grating, was a correct one; and the explanation of the phenomenon was apparent as soon as a careful study of the rotation produced by the various lines had been made.

The absorption lines which we have investigated are numbered, 2, 3, 4, 4', 5, and 6. These numbers conform to those used in

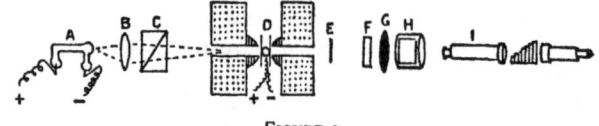

FIGURE 3

A. Mercury arc.
B. Lens forming image of arc on bulb at D.
C. Polarizing prism.
E. Half-wave plate.
F. Cell of bichromate of potash and neodymium to remove yellow and violet lines.
G. Lens forming image of E 3 mm. inside the slit of collimator I.
H. Analyzing nicol.

previous papers on the resonance spectra. The line 4' is much weaker than the others, and did not record itself on the earlier photographs made with vapor less dense than that used in the present case.

The curves of rotation for these lines are shown by Fig. 4: they are only roughly quantitative.

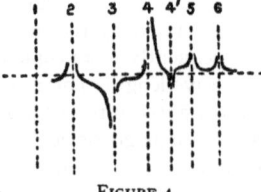

FIGURE 4

The direction of the rotation by the absorption line No. 4 is the same as for the D lines of sodium.

In this group of lines the direction of the rotation changes as we pass from line to line, which explains perfectly why the angular magnitude of the rotation is so small in comparison with that exhibited by the D lines. The $+$ rotation due to a given line is nearly neutralized by the $-$ rotations of its two neighbors. We also see at once why the rotation on one side of a line may be much greater than on the other. For example, we have very strong rotation to the right of line 4 since the opposed rotation due to the faint line 4′ is very small, and the effect contributed by the next line 5 is of the same sign. The rotation to the left of 4 is, however, very small since line 3, with its opposite rotation, lies very close to 4. In the case of line 3 the strong rotation is to the left, since the distance between 2 and 3 is greater than the distance between 3 and 4.

It is very probable that the same condition holds for the vapor of sodium, at least in the green region. In the red and orange portion of the spectrum it is probable that the + and − rotations observed with the Fresnel double prism were due to close groups of lines with rotations of the same sign. It was observed that the luminous needles which penetrated the dark bands were almost invariably found on one side or the other of broad absorption lines, which were undoubtedly unresolved groups of fine lines. Suppose we have a group of a dozen lines, the first four rotating the plane of polarization in the same direction, while for the remainder the sign changes in passing from line to line. It is clear that if the spectroscope does not resolve the lines the Fresnel prism will show a strong rotation in the vicinity of the first lines, that is on one side of the group, and no rotation at all on the other; in other words, we apparently have a broad line which shows rotary power on one side only, which was exactly what was found in the earlier work with sodium.

EXAMINATION FOR THE ZEEMAN EFFECT

Since the selective rotatory power of the vapor in the vicinity of absorption lines can be explained by a longitudinal Zeeman effect, it was of interest to see whether, with the high resolving power at our disposal, any evidence of such an effect could be observed.

We used for the purpose an arrangement employed by one of us in a similar investigation of the vapor of bromine.

A double circular analyzer (two $\frac{\lambda}{4}$ plates of mica, one rotated through 90° with respect to the other) was mounted between the iodine bulb and the analyzing nicol (azimuth 45 with respect to the neutral lines of the plates). One obtains in this way two fields of view separated by a fine line, one corresponding to right-handed, the other to left-handed vibrations.

If a longitudinal Zeeman effect exists, the absorption line, which runs across both fields of view as a continuous line in the absence of the magnetic field, should be displaced in opposite directions in the two fields as soon as the magnet is excited. We were, however, unable to detect a trace of such a shift in the case of any of the iodine lines.

Since the absorption lines 3 and 4 are separated by a distance of about $\frac{1}{20}$ Å. U. we should have certainly been able to detect a shift of 0.01 Å. U.

From this we must suppose, that if the Zeeman effect exists, it is less than 0.01 Å. U. for a field of 20,000 gauss.

RE-ESTABLISHMENT OF LIGHT PERPENDICULAR TO LINES OF FORCE

Cotton has shown that if a sodium flame is placed in a magnetic field, between crossed nicols, the light traversing it in a direction perpendicular to the lines of force is re-established in the vicinity of the D lines, if the planes of the nicols are at 45° with the lines of force. The same phenomenon has been observed by one of us in the case of the non-luminous vapor of metallic sodium.

Voigt and Wiechert have studied the spectrum composition of this re-established light under high dispersion, and have given a theoretical treatment based upon the marked Zeeman effect shown by the lines.

The same experiment has been tried by Cotton with iodine vapor, and by one of us with bromine vapor, with a more powerful field, with negative results.

We have, however, obtained a very marked restitution of light, employing the iodine bulb used in the previous experiments, which was unfortunately too feeble to permit of its examination with the echelon. With the crater of the carbon arc as a source the restored light was quite brilliant, and of the same emerald-green color as in the longitudinal experiment.

DESTRUCTION OF FLUORESCENCE BY MAGNETIC FIELD

Steubing [1] has observed a diminution of the intensity of iodine vapor fluorescence in a magnetic field amounting to as much as 30 per cent. We have repeated the experiment with a much more powerful field and have succeeded in almost completely abolishing the fluorescence.

The effect of the magnetic field in reducing the intensity of the fluorescence becomes more marked as the vapor-pressure of the iodine is diminished.

We found that the form of tube best suited to the study of the phenomenon was as shown in Fig. 5.

A thin-walled tube as free as possible from striæ, and 8 mm. external diameter, is blown out at one end into a small bulb. The tube is highly exhausted and sealed, a crystal of iodine having been introduced before drawing down the tube to a capillary. In exhausting iodine bulbs it is important to cool a portion of the tube leading to the pump with solid CO_2, to prevent the vapor of iodine from entering the pump. After the exhaustion is complete and the tube sealed off, the cooled portion of the tube should be cut away from the pump before the iodine vaporizes.

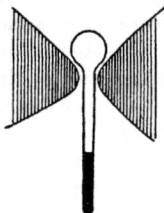

FIG. 5

The tube is mounted between the pole-pieces of the magnet as shown in Fig. 5, and arc or sun light concentrated at its axis with a large lens. Observations are made through the bulb, and it is advantageous to paint the lower portion with black varnish to secure a dark background. The maximum effect is obtained if the lower portion of the tube is cooled to 0° in ice. Under these conditions we estimated the reduction of intensity to amount to fully 90 per cent. with a field of 30,000 gauss, and it is probable that with a field of 50,000 the fluorescence would be practically destroyed. At a tension corresponding to room temperature, the reduction of intensity is much less, and at 35 or 40 degrees scarcely noticeable, though the fluorescence still remains fairly bright in the absence of a magnetic field.

No obvious explanation of the effect of the field in reducing the intensity of and ultimately practically destroying the fluorescence suggests itself. We made numerous experiments to

[1] Ber Deut. phys. Ges., 1913.

determine whether the vapor at very low pressures was thrown out of the field, but these all gave negative results. They were based for the most part upon the principle of allowing iodine to distil from a bulb at 0°, through tubes of the same size into two small bulbs cooled with solid CO_2, one tube passing through an intense magnetic field, the other well outside of it. The phenomenon may result from orientation, but we obtained no evidence of this, though we passed the light through the bulb both parallel and perpendicular to the field and observed the fluorescence in the same way.

The absorption spectrum, as we have stated, shows no change as a result of the field, but it must be remembered that our observations were made with a vapor density corresponding to 30 or 40 degrees, and the reduction in the intensity of the fluorescence is almost imperceptible at this pressure. To observe the absorption at 0° or even at room temperature, it would be necessary to observe with a larger bulb and the field would be less intense as a result.

Conclusions. Up to the present the magnetic rotations of the plane of polarization in the vicinity of absorption bands may be divided into two classes. (1) Anomalous rotations, in which the sign changes in crossing the band, as observed by Cotton for certain solutions and by one of us for a solid film of a neodymium salt. Rotations of this nature appear to obtain in cases in which there is no change in the position of the band of absorption, but merely an alteration in velocities of right- and left-handed circular vibrations.

(2) Rotations in which the sign is the same on opposite sides of the absorption band, as at the D lines of sodium and the iodine lines. Rotations of this type are explained by the division of the line into a Zeeman doublet by the magnetic field.

It seems probable, therefore, that there is a small Zeeman effect for the iodine, but it is doubtful if it can ever be detected as it is of the order of magnitude of the width of the lines, probably much less in fact.

The study of the magnetic rotation of the vapor of sodium by the improved methods outlined in the present paper will undoubtedly give more satisfactory results, as the rotatory power of this vapor in the red and orange is certainly ten times as great as that of iodine.

No. 9

The Fluorescence of Gases Excited by Ultra-Schumann Waves

(In collaboration with G. A. Hemsalech)

An investigation was made a number of years ago by one of the writers, with a view of detecting a possible ultra-violet fluorescence of air excited by waves in the Schumann region.[1]

A small hole was drilled through a plate of aluminum and condenser-sparks discharged against the under side at the perforated spot. If the region above the plate was photographed with a quartz lens in a dark room, it was found that the air above the hole was emitting ultra-violet light, being excited by radiations of some kind which came from the spark. The luminosity had the form of a narrow vertical jet, and its spectrum, photographed with a small quartz spectrograph, showed the so-called 'water-band' of the oxy-hydrogen flame and the ultra-violet bands of nitrogen. The intensity of the radiation was found to be much greater in an atmosphere of nitrogen and much less in one of oxygen. A thin plate of fluorite (1 mm. thick) placed over the hole abolished the phenomenon of the luminous jet completely, from which it was inferred that the excitation was not due to the Schumann waves. It was thought that either luminous molecules were shot out from the spark, or that some sort of corpuscular radiation was responsible for the excitation. The spectrum of the jet was independent of the nature of the metal plate or the lower electrode, and no trace of any of the spark-lines appeared in it, if the gas was free from dust. It was found necessary to exercise great precautions to prevent the formation of dust particles, or nuclei, which scatter the light of the spark, and modify the spectrum of the jet. Metallic dust is given out by the spark, and ultra-violet light causes a cloud to appear in some gases,

[1] WOOD. 'A new Radiant Emission from the Spark', *Philosophical Magazine* [6] vol. xx, p. 707 (October, 1910).

so that when working in closed chambers, there must be a continuous supply of fresh clean gas.

In the winter of 1910 the writers of the present paper commenced an investigation of the subject but came to no very definite conclusions, though some new and interesting phenomena were discovered. It was found, for example, that if a narrow current of air or dry steam was blown across the luminous jet, the luminosity vanished at the spot traversed by the moving air (or steam) current, but retained its full luminosity both above and below the moving gas stream. It was also ascertained beyond any doubt that the luminous material did not come from the spark, for if a stream of CO_2, hydrogen, or coal-gas was directed across the jet, the moving gas current emitted ultra-violet at the point at which it crossed the jet, the spectrum of the emitted light differing in each case. It was also proved that no deviation was produced by a magnetic field.

These results were not published at the time, though they were alluded to in a reply (*Phys. Zeit.* xiii. p. 32, 1912) made by one of us to a criticism by Steubing (*Phys. Zeit.* xii. p. 626, 1911), who claimed to have shown that the jet was nothing but light scattered by dust. His experiments were badly carried out in an apparently hasty manner, and he inferred that, since the 'water-band' can be found in the spectrum of some sparks, the jet must come from the spark, *i.e.*, that it is merely the spark aureole projected through the hole.

Steubing's adverse criticism was quite adequately answered at the time. Most of the results which *he* obtained were undoubtedly due to dust particles, which should have been eliminated.

The photographs illustrating this paper, of the jet in different gases, made with a quartz lens through a quartz prism, show that the spectrum differs according to the gas employed. The spark passed always in air in a closed chamber, and the entrance of the other gas was prevented by keeping an excess pressure in the spark-chamber.

We took up the subject anew during the past autumn, and though we have taken over two hundred photographs, we do not feel that we have accomplished much more than to determine the conditions under which future work must be done. We have, however, discovered many extremely curious phenomena, some of which are still quite inexplicable, but which can be reproduced over and over again with absolute certainty. Most remarkable are the effects obtained with moving and stationary gases: some gases showing a much more brilliant fluorescence when moving across the jet of rays from the spark, others responding vigorously to the excitation when quite stagnant, but showing no luminosity when in motion. The apparatus used in the present investigation is shown in Fig. 1.

It was a box constructed of hard wood in the form of a cross painted black on the inside, and made very nearly gas-tight with

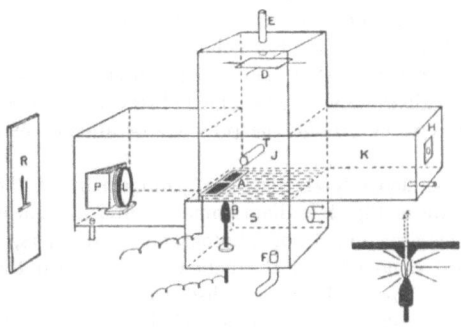

FIGURE 1

putty. The lower compartment S, which will be spoken of as the spark-chamber, communicated with the rest of the box only through the narrow slit at A. This was formed by two jaws of copper about 3 mm. thick, which could be moved to and fro in a brass frame, like the jaws of a spectroscope. The jaws were made of the form shown (in section) by the small diagram of Fig. 1, this form being adopted to insure the sparks striking immediately below the slit. If flat plates are used the spark wanders about, and much longer exposures are necessary. The brass frame and copper jaws formed one electrode, the other (B) consisted of a wedge-shaped piece of copper mounted as shown. The rays of the spark come up through the narrow crevice between the jaws, and are absorbed by the black walls of the upper portion of the box. The other compartments of the box were in free communication, the one to the left containing the quartz-fluorite achromatic lens L, and the 30° quartz prism P, which was cemented over a circular hole in the side of the box. This optical system formed a spectrally dispersed image R on the photographic plate of any luminosity produced in the air or other gas above the slit through which the spark rays passed; the compartment K furnished an absolutely black background against which the luminous jet was photographed. When studying the action of moving gas currents on the jet, the current was delivered against the jet by the tube T. The jet-chamber J was filled with the desired gas through the tube E, a shield D insuring thorough washing out of the upper compartment. The spark-chamber was filled through the tube F. Both chambers were connected with a water manometer, so that the pressure in either could be measured. A small window at H enabled observations to be made of the interior, to determine the presence or absence of a visible cloud.

It will be understood that the arrangement of the photographic plate in Fig. 1 is diagrammatic only, a small extension camera being arranged for photographing the jets, with its front board removed and the bellows attached to the end of the wooden box. The spark discharge was very powerful, and was furnished by a Hemsalech-Tissot resonance transformer with a condenser of 0.05 mfd. capacity. We usually kept the spark-chamber open on one side, ventilating it with an air blast, which prevented overheating of the electrodes; the partition between the two chambers was made of brass, which carried off the heat from the jaws of the slit.

For future work we recommend a metal box, of the same general form as the one which we have used; the chambers K and D are essential, as it is necessary to keep the light of the spark away from the background against which the jet is photographed. The position of the lens L should be arranged so that the spectrum of the light from the jaws of the slit is just barely visible in a dark room with an eyepiece.

Small tubes were instered at the ends of the box to facilitate the washing out of the gases.

The spark-chamber was provided as well with a large hole for the exit of the air, when making experiments with the spark in air. In this way the formation of a cloud of metallic dust was obviated.

It is very important to make sure in each experiment that the 'jet-chamber' is free from dust or smoke which scatters all of the light of the spark. In some gases, initially clean, a cloud speedily forms under the influence of the ultra-violet light, and this must be continuously swept away by a fresh supply of the gas. It was found that the faintest cloud visible to the eye, required an exposure of at least three-quarters of an hour to register itself on the photographic plate; and since most of our exposures were of only twenty minutes duration, we could make sure in every instance, that scattered light played no part in our results.

A quartz prism of 30° was placed in front of the quartz-fluorite achromatic lens used for making the photographs. This spreads out the image of the small point of light formed by the illumination of the walls of the narrow aperture through which the spark rays pass, into the short horizontal spectrum, which appears at the bottom of each one of the figures on Plate V. The rays from the spark pass up in the form of a thin sheet, rendering the gas fluorescent, and the monochromatic images of the luminous sheet of gas are separated by the prism, appearing on the plate as vertical streamers of light.

Similar streamers of course appear if the air or gas in the jet-chamber is charged with dust or smoke, but in this case they are located at the regions of the spectrum where the strongest groups of spark-lines fall.

In the photographs which are reproduced on Plate V the violet region of the spectrum is to the right, the ultra-violet to the left; a small black dot marks the position of the so-called water-band at wave-length 3064. Fig. 1 was made with oxygen in the jet-box, the streamers being due solely to light from the spark scattered by the cloud of small particles which forms almost immediately in this gas. The presence of scattered light (*i.e.* diffused by dust) can always be recognized by a very intense streamer above the violet region of the spectrum (marked V. in Plate V, Figure 1). It will be observed that this streamer is absent in all the other photographs. Figure 2, Plate V, was made with CO_2 in the jet-chamber. Here we have a single streamer practically in coincidence with the left-hand streamer in Figure 1, which is due to the scattering of the strong copper lines 3292 and 3247. Higher dispersion would show this streamer double, the copper lines falling midway between the two bands. Later on we shall refer to the spectrum of the jet in CO_2 made with a quartz spectrograph.

Fig. 3 shows the jet in nitrogen, the spark-chamber being flooded with oxygen during the exposure. The nitrogen was obtained from a bomb of compressed gas, and contained some oxygen. The strongest streamer, immediately above the black dot, is the so-called water-band, while the three bands to the right (position indicated by vertical lines) are the nitrogen bands. This photograph shows us that the gas in which the spark passes is practically without influence upon the intensity or the spectrum of fluorescence of the gas in the jet-chamber. Very remarkable is the jet shown in Fig. 4, also made with nitrogen. In this case the strong streamer to the right appears to coincide with the third nitrogen band (longest wave-length) of Fig. 3, which is very faint in Fig. 3. In fact, it is doubtful if the nitrogen bands of Fig. 3 will appear in the reproduction. Why this streamer is of such great intensity we are unable to say. The previous experiment was, however, made with coal-gas, and we feel inclined to attribute this streamer to some impurity left in the jet-chamber or rubber tube. No trace of the streamer appeared, however, in cases in which the jet-chamber was filled with coal-gas, as appears from Figure 13, Plate VI, in which we have two streamers, one to the right and one to the left of the position occupied by the water-band.

With a current of hydrogen flowing through the jet-chamber (Figure 15, Plate VI), we have a streamer in the same position as the CO_2 streamer. The hydrogen and CO_2 were both obtained from bombs, and it is quite possible that some impurity, such as a volatile constituent of the oil used for lubricating the compression-pumps, may be present in each.

It will be impossible to make any very positive statements until the experiments have been repeated with pure gases.

Figure 11 (Plate V) shows the water-band streamer obtained with a five minutes exposure when the jet-chamber was filled with nitrogen. With the chamber filled with air and an exposure of twenty-five minutes (Fig. 12), we find that the streamer is less intense. The presence of oxygen in the nitrogen may be the cause of the water-band, but if much oxygen is present (as in air) the band is enormously weakened. Strutt has found that the 'afterglow' of nitrogen disappears if oxygen is present, and the fluorescence of iodine excited by ultra-violet light is destroyed also by oxygen.

FLUORESCENCE OF MOVING AND STATIONARY GASES

In our experiments of three years ago an attempt was made to blow the jet to one side by a blast of air, for we were of the opinion at that time that the luminosity might be due to luminous corpuscles projected from the spark, or carried up in the gas blasts projected through the slit by the explosive discharges.

It was found, however, that the air current merely interrupted the jet, the luminosity remaining visible above the moving stream of air. This effect is shown by the photograph reproduced on Plate VI, Figure 14, made with higher dispersion than the figures previously mentioned. The streamer, which is seen distinctly cut in two by the air-current, is that of the so-called 'water-band', the fainter nitrogen bands not showing. The position of the tube delivering the air current is indicated. As we are now quite sure that we are dealing with fluorescence produced by ultra-Schumann waves, it will be necessary to speculate about this phenomena from a corpuscular view-point. The air in the moving current was the same as the air in the jet-chamber, and the experiment establishes the fact that air is fluorescent only when it is stagnant. In other words, it seems as though a given mass of air must be acted upon by the radiations from a number of successive sparks to attain its full luminosity. This appears to be true, however, for the water-bands only. For the nitrogen bands the reverse is true: they are brighter if the gas is in motion.

If a current of nitrogen is directed across the spark jet, the water-band streamer is interrupted and a strong patch of luminosity appears displaced towards one side (Figure 5, Plate V). The direction of the displacement is independent of the direction of the moving current of gas, *i.e.*, it is produced by the prism, as is shown by a comparison of Figures 5 and 6. The jet-chamber was filled with the same nitrogen as that which was flowing in a stream across the jet, yet only the 'water-band' streamer shows below and above the moving gas stream. In Figure 7 the nitrogen was flowing out of the tube at a much higher velocity, and the luminosity is very much less, *i.e.*, we are beginning to get the effect shown by the water-band streamer with lower velocities. The nitrogen in the jet-chamber was evidently set in violent motion by the high velocity current, and the displaced nitrogen bands are visible as complete streamers, instead of being localized in the moving current as in Figures 5 and 6. We next tried blowing a strong current of nitrogen down against the jet, from a tube inserted in the top of the jet-chamber. Its effect is shown by Fig. 8. The water-band streamer is greatly weakened and the nitrogen streamers run up still higher, being almost as intense at the top of the picture where the nitrogen is in rapid motion, as at the bottom where the movement is less violent.

In all cases we found that the time of exposure necessary to record the fluorescent bands of nitrogen could be enormously reduced by keeping the gas in motion. It seems much as if there was a fatigue effect, a flash of fluorescence resulting from the spark radiations, followed by an inability to respond to the radiations from subsequent sparks. It will be remembered that precisely the opposite effect was found for the 'water-bands' which disappear entirely if the gas is in rapid motion. An explanation of these effects will be found in the following paper.

Many experiments have been made to determine the origin of the water-bands. They appeared much stronger in our nitrogen gas than in air, but it must be remembered that our nitrogen contained oxygen. We found, however, that if we added more oxygen to the nitrogen the water-bands were reduced in intensity. If the jet-chamber was filled with oxygen the water-band streamer was exceedingly faint, perhaps $\frac{1}{20}$ of the intensity which it had in the case of the nitrogen, and no other streamers were visible. A current of coal-gas gave the double band shown in Figure 9.

If a moving stream of oxygen is directed across the jet, the jet-chamber being open on both sides and filled with air, that portion of the water-band streamer which is crossed by the oxygen stream disappears, while a displaced band appears in the same position as the nitrogen band, only much fainter (Figure 10, Plate V). We might attribute this to an impurity of nitrogen in the oxygen, brought out by having the gas in motion, but we must remember that a moving current of air gives no trace of any displaced band. It is quite probable that impurities present in the gases or derived from the rubber tubes through which they passed are responsible for many of the apparently inexplicable phenomena found in this preliminary investigation.

It appears to be pretty definitely established that oxygen has a tendency to destroy the fluorescence of gases with which it is mixed. This accounts for the extreme faintness of the water-band streamer in oxygen. The oxygen molecules destroy their own fluorescence, so as to speak, the phenomenon being analogous to the one observed by one of us in the case of iodine.[1]

A study of the fluorescence of iodine vapor under the influence of the spark rays has been commenced by one of the writers and will be reported in a subsequent paper.

Delivered in a stream of warm nitrogen across the spark-jet it fluoresced with a bluish-green light, while if a current of warm *air* was used there was no trace of any fluorescence. If a thin plate of quartz was placed over the slit the fluorescence was visible for a few seconds and then rapidly faded away. If, however, the quartz plate was moved a little, the fluorescence appeared again. It was found that the opacity of the quartz was produced by an almost imperceptible film deposited upon it by the spark.

We had made many experiments with thin plates of quartz and fluorite, but had failed to find any trace of the water-band streamer, though we obtained faint indications of the coal-gas streamer and possibly those of CO_2.

The observations made with iodine threw a new light on the subject and made it necessary to repeat the experiments under conditions which would preclude the formation of the opaque deposit. This was accomplished by moving the quartz plate about during the exposure. Using this precaution the nitrogen streamer was obtained by giving an exposure of only five minutes. Its intensity was equal to that obtained without the quartz plate with an exposure of one minute. No trace of the water-band streamer was found, however.

[1] Wood and Speas. 'A Photometric Study of the Fluorescence of Iodine Vapor', page 77. This monograph.

A fluorite plate 1 mm. thick was next tried, and a faint but unmistakable image of the water-band streamer was obtained with a fifteen minutes exposure. Without the fluorite plate a stronger image of the streamer appeared with a one-minute exposure. A rough estimate indicates that the fluorite plate reduces the intensity of the radiations which are responsible for the excitation of the water-band to about 5 per cent. of their original value. This circumstance makes it seem extremely probable that we are dealing with ultra-violet waves much shorter than the Schumann waves, which pass readily through fluorite lenses and prisms. The radiations which excite the nitrogen bands appear to be reduced to about 20 per cent. of their original intensity after passing through a quartz plate 1 mm. in thickness. Very few experiments have been made with the plates, and the apparatus was not very well adapted to the work, as it is necessary to keep the plates in motion or clean them every ten or fifteen seconds. With a properly designed apparatus it is probable that much more reliable data can be secured. It will also be possible to work with gases at a low pressure, in which case the fluorescence is likely to be very much brighter, judging from the behavior of iodine, which ceases to fluoresce under the stimulation of visible light when at a pressure of a few millimeters only, its greatest luminosity appearing when it is at a pressure of about 0.2 mm.

A number of photographs of the jet were made with a quartz spectrograph of medium size, an exposure of several hours being necessary. Figure 16, Plate VI shows the spectrum of the jet in nitrogen with a current of nitrogen delivered across the slit. The reduction of the intensity of the water-band, where it is crossed by the nitrogen stream, is very marked. The line to the left of the water-band is the head of the nitrogen band at wave-length 3159. It falls within the less refrangible and fainter portion of the water-band; (see Fig. 18 for complete water-band) then come the two strong copper lines 3292 and 3247 (in the continuous spectrum of the light reflected from the slit jaws), then the nitrogen band 3369 (A), and the bands 3527–3576 (B)—3755–3802 (C)—3914 (D).

Fig. 19 was made with a hydrogen current crossing the jet. A faint continuous band appears above the strong copper lines (which show also probably as the result of the formation of a slight trace of fog or cloud, which sometimes happens if a very slow gas-current is used). The water-band is very well shown in this photograph, and can be detected above the region where it is interrupted by the hydrogen stream. In making this experiment the side of the box was opened and the hydrogen allowed to escape without filling the jet-chamber.

Fig. 17 shows the spectrum obtained with a current of CO_2 blowing across the jet. It appears to be identical with the band found with hydrogen, so far as its position is concerned, but on plates made three years ago the distribution of intensity in the two bands was quite different.

No. 10

A Further Study of the Fluorescence Produced by Ultra-Schumann Rays

(In collaboration with C. F. Meyer)

In 1910 one of the writers[1] described experiments showing the existence of a radiant emission from the spark which had not been previously detected. The subject was more fully investigated in collaboration with G. A. Hemsalech[2], and many interesting phenomena observed. All experiments indicated that the emission which was being studied consisted of ultra-Schumann waves. No method was found, however, of determining their wave-length, and the experimental difficulties throughout the work were so great that many phenomena were only incompletely studied, and many points were left uncertain. The present authors have therefore attacked the problem anew, and the results they have obtained will be discussed in this paper.

The radiant emission in question cannot be directly observed or photographed, but its existence is shown by the fluorescence which it causes in certain gases. The essential parts of the apparatus used in its study were the same as in the above-mentioned investigations, except that the box forming the jet-chamber this time consisted of metal instead of wood.

APPARATUS

Referring to Fig. 1, A is a circular copper plate 3 mm. thick and 7.5 cm. in diameter, part of the plate being represented in the diagram as cut away. The plate rested on and was sealed to a short piece of brass tubing 7 cm. in diameter and 1 cm. long, which in turn was soldered over an opening of nearly as great diameter in the bottom of the metal box J. Through a hole in the center of this plate a truncated portion of a copper rivet had been

---

[1] Wood, *Philosophical Magazine.* [6] xx. p. 707 (1910).
[2] Wood and Hemsalech, *Philosophical Magazine* [6] xxvii, p. 899 (1914).

driven. The rivet had a vertical slit S cut into it, 2 mm. long and .2 mm. wide. B is a piece of heavy copper wire. The copper rivet in the center of plate A and the wire B served as terminals between which a spark from a transformer was passed. The radiation from the spark passed up through the slit S into the jet-chamber J, together with the ordinary visible and ultra-violet light. The special radiation, which we may call ultra-Schumann radiation, causes a short jet of ultra-violet fluorescence in the air or other gas above the slit S in the jet-chamber J. This fluorescence was photographed through the quartz window W_1 by

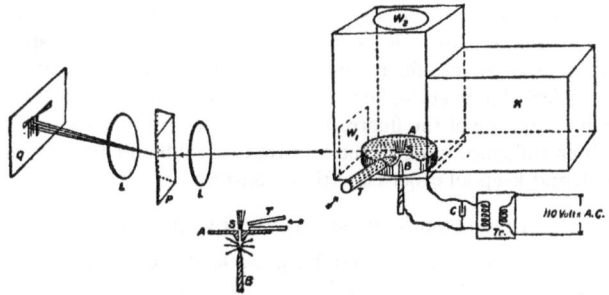

FIGURE 1

means of the quartz spectroscope represented diagrammatically by the lenses L, the prism P, and the photographic plate Q. The spectroscope had had its slit-tube removed, the jet itself serving as slit. The auxiliary chamber K is added to the apparatus in order to ensure a dark background. The chambers J and K (13 × 9 × 9 cm.), which consisted of metal boxes blackened inside, served as a dark enclosure, and also as a container for the particular gas in which it might be desired to study the jet.

The tube T served for the introduction of the desired gas, and was so placed that the stream delivered by it passed directly over the slits, *i. e.*, through the fluorescent jet.

The arrangement of the plate A, slit S, lower electrode B, and the tube T can also be seen in the small sectional diagram of Fig. 1, which is drawn as seen by an observer looking from the direction of the spectroscope.

The glass window W_2 was introduced for the purpose of determining whether the radiations from the spark were coming up through the slit properly.

During some of the experimental work, all lines along which parts of the apparatus joined were sealed up with wax, and the bottom of the box J was sealed with mercury. During other parts of the work the lines of juncture were left unsealed.

The spark was produced by a ½ kilowatt, 110 volt, 60 cycle transformer. Across the spark-gap was placed a condenser consisting of 36 copper plates, 15 × 20 cm., insulated by ordinary window-glass, the whole being immersed in oil. No influence of the nature of the spark upon the nature of the fluorescent spectrum of the jet was ever ascertained, except that a strong spark gives a more intense fluorescence than a weak one. The authors do not feel able to state, however, that there is no influence at all upon the nature of the fluorescence, the difficulty of ascertaining such an influence lying in the fact that it is difficult to vary the spark and keep all other conditions constant.

THE FLUORESCENT SPECTRUM IN VARIOUS GASES

In the work of Wood and Hemsalech above referred to, it was found that the strongest fluorescence of the jet was obtained when nitrogen was used in the jet chamber, the spectrum then consisting of the water band λ 3064 Å U., and under favorable circumstances the nitrogen bands. Their nitrogen, however, contained impurities, especially oxygen; and the present authors thought it to be of interest to determine the effect of removing the last trace of oxygen. Nitrogen obtained from a bomb was accordingly cleared of the oxygen it contained by the method described by C. Van Brunt,[3] dried, and passed into the jet-chamber. The spectrum showed the water-band and to the left of it the nitrogen bands. In the commercial nitrogen only the first and second, counting from the water-band, are present; while in the purified nitrogen the first (3369) is faint, and the second and third (3556 and 3778) are prominent. Some time after these photographs were obtained, when the apparatus had all been taken down, cleaned, and set up again, a different type of spectrum was obtained with purified nitrogen, in which the

[3] *Journal of the American Chemical Society*, July, 1914.

first and second nitrogen bands were faint and the third strong. The type of spectrum in the commercial nitrogen remained the same. The cause of the difference was not discovered even after making a number of exposures under various conditions for the purpose of ascertaining it.

Some little time was spent in repeating the more important parts of the work of Wood and Hemsalech, done in Paris. With entirely new apparatus, and different sources of supply of our gases, these attempts at repetition often resulted at first in very perplexing failures. Investigation into the cause of these preliminary failures, however, resulted in most cases in throwing much light on the phenomena themselves. For example, the first attempts to duplicate the photographs obtained with moving and stationary gases were unsuccessful. In these a stream of nitrogen was blown across the fluorescent jet, the stream causing an interruption of the water-band, and the appearance of the nitrogen bands in the moving stream. Later attempts showed that in order to obtain the interruption of the water-band it was best to bring the mouth of the tube T close up to the slit S (say within 0.5 cm.). Experiments made with smoke indicated, however, that the stream of gas delivered by the tube at the velocities which we were using maintained approximately its form and area of cross-section, and consequently its velocity, for several centimeters beyond the mouth of the tube. This suggested that the appearance of the water-band in the stationary portion of the gas, and its disappearance in the moving stream, might be due, not primarily to the rest or motion of the gas with reference to the fluorescent jet, but to a difference in the constitution of the gas just leaving the tube T, and that immediately around the stream, even though the jet-chamber was entirely closed, and had been washed out with a stream of nitrogen delivering a liter and a half per minute for four minutes before beginning the exposure.

To increase the difference in constitution between the gas in the stream and that surrounding it, an opening was made in the jet-chamber to admit air. Purified nitrogen was blown across the slit and an exposure made. It was found that the water band was again interrupted where the nitrogen jet crossed it, and that the nitrogen bands appeared only in the stream, which proves that the smallest trace of oxygen prevents the appearance of the nitrogen band.

An attempt was also made to explain the interruption of the water-band found when a stream of air was blown over the slit S, and the jet-chamber was filled with air, as being due to a residual difference between the air in the stream and that in the jet-chamber. When air from the room, which of course also filled our apparatus, was blown by means of a bellows through the tube T over the slit, no interruption of the water-band was shown. When the air was moistened by passing through wet cotton still no interruption was shown; but when the air from the bellows was dried and blown over the slit, and a source of moisture was provided in the jet-chamber, so that the stagnant air around the air current might take up moisture, the photograph showed the water-band interrupted by the jet. Moreover, when a strong current of dried air (about 2.5 liters per minute) was blown into the apparatus for five minutes before exposure was begun, and no source of moisture was provided in the jet-chamber, the fluorescence was so faint that it could not be photographed in fifteen minutes, which is about the time of exposure of the other photographs. It thus appears that water vapor is necessary to obtain the fluorescence of the water-band, while the presence of oxygen mixed with nitrogen, either in large or small quantities, will not give it.

An exposure was also made in which nitrogen purified and subsequently moistened was blown across the slit S. This plate shows the water band uninterrupted where the current of nitrogen crosses the fluorescent jet; shows faintly two of the three nitrogen bands 3369 and 3556 on the long wave-length side of the water band, and a fourth band or line on the short wave-length side, probably the fainter water band 2811. The fact that the water band is interrupted when the stream of nitrogen delivered by the tube T is perfectly dry, but is no longer interrupted if the nitrogen is moistened, would lead us to believe that if the stagnant gas in the jet-chamber consisted of entirely pure dry nitrogen, then the fluorescent spectrum would consist of the three nitrogen bands only.

Experiments tried with gases other than nitrogen and air were not carried far enough to lead to results of sufficient interest and certainty to warrant discussion, except in the case of iodine vapor. Some crystals of iodine were placed in a glass tube through which nitrogen was passed, and the resulting mixture of iodine vapor and nitrogen was blown from the tube T across the slit. The iodine fluoresces in the visible region with a bluish-green light as was noted by Wood and Hemsalech, and in the ultra-violet. This fluorescence was so bright that it could be easily photographed by throwing an image of the fluorescent jet upon the slit of a spectroscope with a quartz lens, thus obtaining its spectrum in very much greater detail. This was done for us by Mr. Voss of this laboratory, with an improvised quartz spectrograph of the Littrow type, furnished with two Cornu prisms and a lens of about two meters focus. A strong narrow band appeared in the ultra-violet made up of lines as follows:

3379.7	3414.8
3388.8	3418.5
3400.0	3423.9
3406.3	3426.4
3408.8	3431.4
3413.3	3435.3

The spectrum of the visible bluish-green light has not yet been photographed, as it is very much fainter than the ultra-violet band.

TRANSMISSION, REFRACTION, AND REFLEXION

In the study of the transmission of the rays exciting the fluorescent jet an entirely new jet-apparatus was used, the spectroscope and electrical equipment remaining the same. This new apparatus was based on the same general principles as the old one, but differed in being smaller; in providing for the delivery of the fresh gas from the rear through a large tube (1.5 cm. in diameter), which at the same time formed a dark background for the jet; in having a special mounting for the truncated copper rivet with the slit forming the one electrode, and the copper wire forming the other, so that these could be readily removed and replaced to allow for cleaning of the plate the transmission of which was being tested. Moreover, the entire jet-apparatus was mounted in a position inverted with respect to that shown in Fig. 1, so that the spark was above, while the jet pointed down. The new form and mounting are represented diagrammatically in Fig. 2 in the text. J is again the jet-chamber, T the tube delivering the gas, also, serving as dark background, W_1 the window through which the jet is photographed, R the rivet having a slit in it in the plane of the drawing, B the wire forming the other electrode. R is driven through a thin copper plate A; A and B are both fastened to an arm which rotates about a horizontal axis in the plane of the drawing, so that both may be readily raised, by raising this arm, and then replaced.

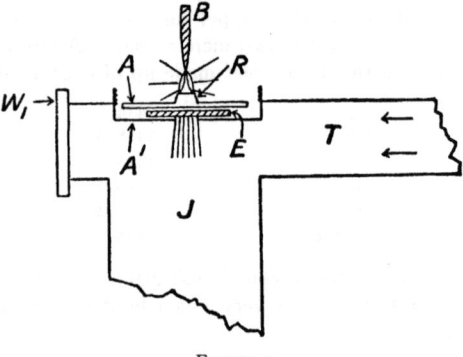

FIGURE 2

E is the plate or lamina whose transparency is to be tested; A' is a thin brass plate upon which E rests, and serves also to prevent light scattered by the edges and surfaces of E from reaching the spectroscope. The spectroscope is on the left as in Fig. 1.

The first experiments in transmission were made with a small disk of clear colorless fluorite 0.57 mm. thick. The photograph showed the transmission of the rays exciting fluorescence of the water-band. The gas passing into the jet-chamber through the tube T consisted of nitrogen taken directly from the bomb, and the fluorescent spectrum when the fluorite plate was not in the apparatus consisted of one very broad streamer indicating the water-band and the one or two nitrogen bands nearest it all merging into each other. When the fluorite plate was inserted only the water-band appeared, showing that the fluorite was more transparent to the radiation exciting the water-band than to that exciting the nitrogen bands. Three exposures made under approximately the same conditions indicated that an exposure of from fifty to seventy times as long was necessary to obtain an image of the same intensity for the water-band when the fluorite plate was in as when it was absent. This indicates that the radiation lies outside of the so-called Schumann region, for which fluorite is quite transparent. In making these exposures it was necessary to clean the fluorite plate frequently. It was as a rule wiped with dry cotton every two minutes, and every fifteen or twenty minutes cleaned with nitric acid, water, and alcohol. The time taken for wiping and cleaning is of course not included in the time of exposure. In these exposures the amount of nitrogen delivered into the jet-chamber was about a quarter of a liter a minute, the apparatus being washed out for from five to ten minutes before the exposure was begun.

When purified nitrogen was used the fluorescent spectrum showed the nitrogen band of longest wave-length predominant. Inserting the fluorite plate into the apparatus showed that the radiation exciting this band was also transmitted, but very much less readily than the radiation exciting the water band—perhaps a third or fifth as well.

The study of the transmission of quartz was attended by more difficulties, and yielded less definite results than that of fluorite. Using nitrogen directly from the bomb, in the spectrum of which the water band predominated, no transmission of the radiation exciting the water band has ever been detected, though the conditions under which the exposures were made were not entirely favorable. Using purified nitrogen, in which the nitrogen band of longest wave-length came out most strongly, it was found that after transmission through quartz the region of most intense fluorescence was displaced toward the short wave-length side of the spectrum, indicating that it was probably the radiation exciting one of the nitrogen bands of shorter wave-length which was being transmitted. In all the photographs of long exposure when purified nitrogen was used, there is also a fluorescence at about 2300 ÅU. The radiation exciting this fluorescence is not very strongly absorbed by either quartz or fluorite. Exposures made to detect a possible transmission of any of the radiation exciting fluorescence through laminæ of fused quartz and thin films of mica gave entirely negative results.

If the radiation from the spark which has been studied in this paper is really of the nature of light, then it should be possible to refract it by passing it through the thin edge of a prism of quartz or fluorite. A quartz prism of about 10°, ground to a razor edge by Petitdidier was accordingly placed in the apparatus in such a way that the radiations passed through the knife-edge of the prism. The edge of the prism was pressed against a piece of black paper to prevent any of the radiation from passing by

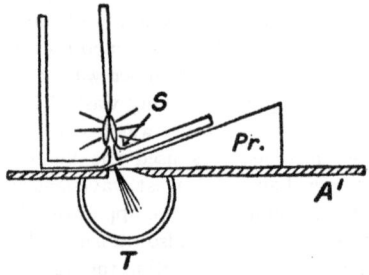

FIGURE 3

without going through the quartz. For this experiment the original jet-apparatus (Figure 1) was used, mounted in inverted fashion and adapted to the present work. A tube was provided to feed the nitrogen from the rear (similar to the tube T, Figure 2). The arrangement of the spark, slit S, prism Pr, bedplate A', and tube T viewed end-on, is shown in the sectional diagram, Figure 3, in the text, which is drawn as seen from the direction of the spectroscope.

Passing a current of purified nitrogen of about one-third liter per minute into the jet-chamber and exposing for an hour and a quarter, cleaning the prism frequently, we obtained a photograph in which the streamers were inclined at an angle showing that the radiation is strongly refracted. Further, since the fluorescent streamer is fairly broad, it indicates that for an extremely thin refracting edge as was here used, the rays exciting at least two, and possibly three of the nitrogen bands, and possibly also those exciting the water band, are transmitted through the quartz. In the original photograph the fluorescent jet of wavelength about 2300 ÅU., mentioned above as being present in long exposures when purified nitrogen is used, was also seen.

To obtain a value of the deviation of the rays by the prism in the jet-apparatus several exposures were made, using the spectroscope merely as a camera (*i. e.* with the prism of the spectroscope removed). An exposure was then made with smoke in the jet chamber. The light scattered by the smoke was filtered through a silver film, and thus the deviation produced by the prism in the jet-apparatus upon light of wave-length 3000 ÅU. (the transmission band of silver) was obtained. The index of refraction of quartz for this wave-length is given in tables as 1.57. From this value and the relative deviation for this wave-length and for the light exciting the fluorescence, we calculate for the latter an index of $1.75 \pm .08$, that is, a probable value somewhat greater than the greatest index in the ordinary transmission region of quartz. The object of finding the index of refraction of quartz for these rays was to see if it might not be less than the index for ordinary ultra-violet light. If this had proven to be so, it would have indicated that the rays exciting the fluorescence were on the short wave-length side of the ultra-violet absorption band of quartz. But as this has not been found to be true, we must conclude that they are still on the long wave-length side of the absorption region, unless indeed they should be on the long wave-length side of an entirely hypothetical second quartz absorption band, still further out in the ultra-violet. Experiments under way show that the radiation exciting the water band can also be reflected.[1] The reflecting surface used was a cathode deposit of silicon on glass made by Dr. E. O. Hulburt. At 45° incidence the amount of reflexion lies between ten and twenty per cent.

[1] Recent experiments by Dr. Meyer (*Physical Review*, 1917) show that the probable wave-length of the radiation is about 1300 ÅU. He mounted a small grating of speculum metal with 30,000 lines to the inch over the slit and obtained streamers due to the regularly reflected pencil and the diffracted pencils of the first and second order spectra. The fluorescence may be excited in part by rays of this wave-length and in part by rays further down the spectrum which are not readily transmitted by fluorite, or reflected by speculum metal.

The fluorescent spectra excited by ultra-Schumann radiation in nitrogen containing some oxygen, in nitrogen free from oxygen, in moistened nitrogen, in dry and moist air, and in nitrogen with an admixture of iodine vapor, have been studied. An explanation for the effects obtained with moving streams of gas is given.

Fluorite and crystalline quartz are slightly transparent for the rays studied. Those exciting the fluorescence of the water band pass more easily through fluorite than do those exciting the nitrogen bands. For crystalline quartz the reverse is probably true. Rays exciting a very faint fluorescence at about 2300 ÅU. in purified nitrogen pass readily through fluorite and crystalline quartz. Fused quartz and mica are found opaque for all the rays exciting fluorescence.

The ultra-Schumann radiations have been refracted. The index of refraction of quartz for them is $1.75 \pm .08$, indicating that they lie on the long wave-length side of the quartz absorption band.

The radiation exciting the water-band has been reflected.

No. 11

Scattering and Regular Reflection of Light by an Absorbing Gas

In 'Researches in Physical Optics', Part I, an account was given of a series of experiments on the selective reflection, scattering and absorption by resonating gas molecules.

In this paper it was shown that mercury vapor at the pressure which it has at room temperature (about 0.001 mm.) in an exhausted quartz vessel, when illuminated by the light of a quartz mercury arc, re-emits diffusively a highly monochromatic radiation, of wave-length 2536, as a result of the presence in the illuminating beam, of rays of the same wave-length.

Energy re-emitted in this way without change of wave-length, by resonating molecules has been named Resonance Radiation, and was first observed by the writer, in the case of sodium vapor, in 1905.[1]

Since the appearance of my first paper on the remarkable optical phenomena exhibited by mercury vapor, a number of papers, by various authors, have appeared, dealing with the same subject, and still more recently, in collaboration with M. Kimura, I have investigated and cleared up a number of doubtful points and outstanding problems.

To make the present paper a complete statement of what is known on the subject at the present, it will be necessary, not only to refer to the work of the other investigators who have been occupied with the subject, but also to incorporate some of the results obtained in the earlier investigation. The fundamental experiment was made in the following way. A small drop of mercury was introduced into a tube of fused quartz, closed by end plates of the same substance which had been ground flat

[1] R. W. Wood. Fluorescence of Sodium Vapor, and Resonance Radiation of Electrons. *Philosophical Magazine*, November, 1905.

and polished. These plates were fused to the ends of the tube, which had been flared out in order to prevent spoiling the figure of the central portion of the plates by fusion. The tube was highly exhausted and sealed, and the light of a quartz mercury arc focused along the axis of the tube. The tube was now photographed from the side with a camera furnished with a lens of quartz, which was constructed in a few minutes from an old box used for storing photographic negatives. The photograph showed an image of the cone of rays traversing the high vacuum precisely as if the tube were filled with dense smoke. The tube was at room temperature, and the density of the mercury vapor was about 0.001 mm., nevertheless an exposure of fifteen or twenty seconds was all that was necessary.

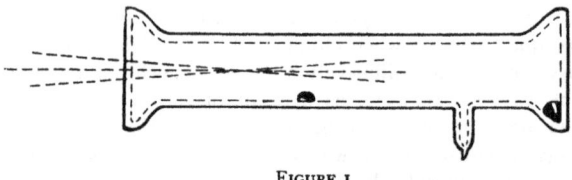

FIGURE 1

The first experiment was made with an old Heraeus quartz lamp operated cold with the current from a high potential transformer.

On repeating the experiment subsequently with a powerful Westinghouse quartz arc, no trace of the phenomenon was observed. It occurred to me, however, that in the present case I was working with the lamp at high temperature, and that the 2536 line was very likely reversed, the wave-length necessary for the excitation of the resonance radiation being removed by absorption. I accordingly allowed the lamp to become quite cold, and made my exposure when the lamp was first lighted, when its light is quite violet in color. On developing the plate I found that a five seconds exposure gave me a more intense cone of light than anything that I had ever observed before. It was absolutely black on the negative. Moreover, the vapor outside of the cone of focused rays appeared to be glowing at the end of the tube where the beam passed in. This photograph is reproduced on Plate VII, Figure 2. A photograph of the spectrum of the light emitted by the vapor showed only the 2536 line, though the vapor was illuminated by the total radiation of the lamp. Figure 7, Plate VII, upper spectrum resonance radiation, lower spectrum mercury.

In most of the subsequent experiments the exciting radiation was restricted to the 2536 rays by first passing the light through a quartz monochromator, and then focussing the emergent rays on the exhausted quartz vessel.

If the temperature of the quartz tube was gradually raised, it was found that the luminous cone became shorter and brighter, until it finally disappeared, the emitted light coming from the inner surface of the plate where the exciting radiation entered. This diffuse resonance radiation of mercury vapor at different densities has been reinvestigated in collaboration with M. Kimura. In our work we have used a water-cooled quartz mercury arc of the type described by Kerschbaum: (*Electrician*, London, 1914, vol. 72, p. 1074), in which the arc is driven against the front wall of the tube by a weak magnetic field. This reduces the self reversal to a minimum, for the cooler, non-luminous absorbing layer is 'squeezed out' so to speak, the current-carrying vapor being in contact with the quartz wall of the tube. The spectrum of such a lamp is quite unique in appearance, for the 2536 line is so much brighter than any of the other lines that it is enormously over-exposed, appearing much like a photograph of a distant arc light taken at night. Bright diffraction rays radiate from it in all directions, causing it to stand out on the photograph with the conspicuousness of a first magnitude star on the milky way. A photograph of the spectrum is reproduced as a negative on Plate VII, Figure 8. The wavy lines joining the two spectra were caused by the elevation of the plate between the two exposures. The lamp consists of a straight tube of quartz with the negative electrode (mercury) below, and a positive electrode of tungsten above. It was made to order by the Cooper-Hewitt Co. and was mounted in a brass water jacket as shown in Figure 2. Two brass tubes, slightly larger than the lamp tube were soldered into a brass box, made by boring out a solid cube of brass. The hole bored through the side of the cube was covered with a circular plate of crystalline quartz cemented to the brass with a mixture of bees-wax and rosin. The lamp was started by tipping it on its trunions of soft iron, which were magnetized by coils in circuit with the lamp. The coils are of course connected in such a way as to produce a magnetic field in the box directed so as to deflect the arc against the side of the tube nearest the quartz window.

The lamp operates on 110 volts, with resistance sufficient to hold the current down to about 3.5 amperes. During operation

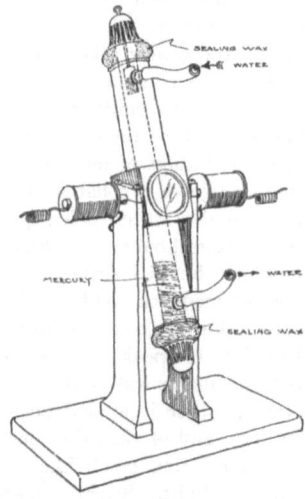

FIGURE 2

the anode is at a full red heat, but the rapid circulation of water in the brass jacket keeps the tube quite cold. The light of the quartz arc was passed through a quartz monochromator arranged to give a convergent cone of 2536 monochromatic light. It was simply a roughly constructed quartz spectroscope with a very wide slit and no telescope tube, shown in diagram in Figure 4. The image of the slit formed by the 2536 rays was located in space by means of a strip of uranium glass, and the bulb mounted in such a position that the image fell upon the center of the prismatic plate. The dispersion was sufficient to remove the other images of the slit from the bulb, which obviated the use of a second slit and lens for obtaining the monochromatic illuminating beam. For investigating the resonance radiation at different densities we used a thick-walled bulb of fused quartz, with the bottom made of a slightly prismatic plate of optical fused quartz. (Figure 3.) The plate was made prismatic so that the rays reflected from

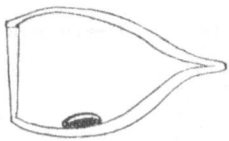

FIGURE 3

the inner surface could be studied uncontaminated by rays reflected from the outer surface. The bulb was mounted over a chimney of thin sheet iron, with a Bunsen burner at its base and the temperature determined by a nitrogen filled mercury thermometer, with its bulb in contact with the upper surface of the bulb. A camera of very simple construction, furnished with a quartz lens was focused upon the bulb, the process consisting in first focussing it with uranium glass upon the image in space

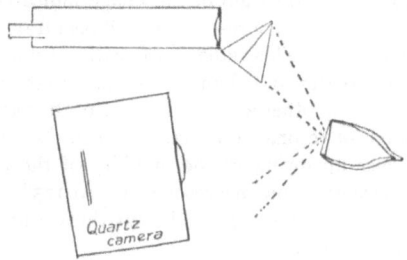

FIGURE 4

of the 2536 line formed by the monochromator, and then measuring the distance between the lens and the image in space. This gives us the proper distance of the bulb from the lens to secure a sharp focus.

The arrangement of the apparatus in this experiment is shown in Figure 4, the rays reflected from the two surfaces of the prismatic plate falling to one side of the lens. In this way we obtain only an image of the scattered resonance radiation from the bulb.

The photographs obtained at different temperatures are reproduced on Plate VII, Figure 5.

The temperatures and resulting pressures of the mercury vapor are as follows:

Temp.	Pressure of Hg. Vapor.
A 23°	.00168
B 40°	.00574
C 76°	.0750
D 100°	.276
E 175°	11.00

The bulb at room temperature (A) appears more or less filled with the resonance radiation. This, as has been shown previously, is due to the fact that the radiation from the molecules which lie in the path of the primary beam excite to resonance the entire mass of vapor in the bulb. At 40° (B) the pressure has increased about 3.4 times, and the radiation comes chiefly from the front part of the bulb.

At 76° it is confined chiefly to the inner surface of the plate, though a slight haze to the right of the image indicates that some radiation capable of exciting resonance, still penetrates to a depth of a millimeter or so. This radiation is without doubt of wavelengths slightly greater and slightly less than that of the center of the exciting line, in other words, the edges of the 2536 line. The size of the patch of resonance radiation contracts rapidly as the temperature goes up owing to the inability of the radiation to spread out and excite secondary resonance. At 175° it has shrunk to the dimensions of the image of the slit thrown upon the bulb by the monochromator.

It will be noticed that in case (D) where the density is 16 times as great as at room temperature there is still a slight broadening of the image due to the secondary resonance. The spreading is of the order of half a millimeter which is about what we should expect from the known stopping power of the vapor at room temperature. A further increase of temperature causes a rapid diminution in the intensity of the scattered resonance radiation, the energy of the primary beam passing off as a regularly reflected wave. The intensity is a maximum in the vicinity of 100° (*i.e.*, at a pressure of about 0.3 mm.). At 150° (pressure about 3 mm.) the intensity has decreased to about half its maximum value; at 200° (pressure 18 mm.) to about one-quarter, and at 250° (pressure 76 mm.) to perhaps one-tenth. At 270° there is absolutely no trace of the scattered radiation. The estimates were made from a series taken under similar conditions, with a slit somewhat broader, so that a better determination of the relative densities could be made. Equal exposure times (40 secs.) were given and the images developed simultaneously. It is evident that the scattered resonance radiation decreases (replaced by true absorption probably) long before regular reflection commences.

It was found, in the earlier work, that the admission of air introduced the factor of true absorption, and there is no doubt but what an increase in the density of the mercury vapor, in the absence of air, acts in the same manner.

A related phenomenon occurs in the case of iodine vapor, the fluorescence of which reaches a maximum value when the pressure is about 0.25 mm. and drops to one-half of this value at a pressure of 1 mm., as was found in an earlier investigation in collaboration with W. P. Speas, included in this monograph.) From results obtained in collaboration with J. Franck we now know that the destruction of the fluorescence of a gas or vapor by a chemically inert gas, depends chiefly on the electro-negative quality of the gas which is introduced. The effect is least with the gases of the helium group, and greatest with strongly electro-negative gases, such as carbon dioxide and chlorine. In the case of the visible (bluish-green) fluorescence of mercury vapor, excited by light from a spark, it was found that the emission persisted up to and above atmospheric pressure if the pressure resulted from mercury vapor alone, whereas the admission of air at a few centimeters pressure, destroyed it entirely.

It thus appears that the ultra-violet resonance radiation is more easily destroyed than the visible fluorescence, by an increment in the number of mercury molecules in unit volume.

In the earlier work it was found that if the density of the mercury vapor was sufficiently increased the scattered resonance radiation was replaced by selective reflection, and the very attractive hypothesis was made that the secondary wavelets, emitted by the resonating molecules, combined by Huygen's principle to form a regularly reflected wave. In the light of more recent experiments it appears that we must use caution in adopting this hypothesis, since it appears that the diffuse radiation disappears entirely some time before regular reflection commences. This makes it seem probable that the mechanism by which reflection occurs is more nearly such as we have in the case of metals, except that in this case it is very sharply selective.

At all events it seems certain that we must regard the resonators as quiescent when the density is sufficient to give rise to a regularly reflected wave. There, is of course, still a possibility that the reflected wave may be regarded as built up of the wavelets coming from the resonators, for an amount of emission too feeble to detect when diffusive, would give considerable illumination when gathered into a regularly reflected wave.

The fact, however, that practically all trace of diffuse radiation disappears at 3 cms. pressure, while selective reflection appears at about 12 cms., reaching its maximum at perhaps one or two atmospheres, makes it seem doubtful whether we can regard the reflected wave as built up by the wavelets emitted by the resonators. Moreover, experiments indicate that the light emitted by the resonators under polarized stimulation, is *unpolarized*, while the regularly reflected ray is completely polarized. This matter will be considered again further on.

SELECTIVE REFLECTION

The selective reflection of mercury vapor was first described by one of us in the *Philosophical Magazine* for July, 1909.

The earlier investigation of the transition from diffuse scattering to regular reflection was made with a spherical bulb of fused quartz. This was not well adapted to the work, as the reflected beams diverged from the small virtual images of the source formed by the spherical surface and experiments with polarized light were impossible. In the present work we have employed the bulb closed with the prismatic plate already described. It was supported in a horizontal position by a double loop of nickel wire and turned so that the prismatic plate threw off the two reflected beams to the side. In the first experiment it was placed a little inside of the focus of the monochromator, so that the incident radiations came to a focus after reflection from the prismatic plate. A plate of uranium glass was mounted in such a position that the two reflected images were focused on it. The image formed by reflection from the outer surface was noticeably brighter than the other, owing to absorption by the fused quartz plate, which was twice traversed by the rays reflected from its inner surface. On heating the bulb to a red heat with a Bunsen flame, the latter image brightened up until it appeared to be about three times as bright as the image reflected from the outer surface. In this way it is possible to demonstrate the selective reflection of the vapor to a small audience at close range. A sheet of heavy p'ate glass must be used as a protection against a possible explosion of the bulb as the pressure may rise to 15 or 20 atmospheres.

The reflecting power was next determined quantitatively in the following way:

The total radiation from the water-cooled arc was reflected from the inner surface of the prismatic plate into a small quartz spectrograph, the slit of which was opened rather wide, and shortened to a length of about 1 mm. The spectrum lines thus photographed as small rectangular patches, and twenty or thirty exposures could be made on a single plate. The exposures were made by a slow swinging shutter of the pendulum type. We first made a series of exposures of continuously increasing duration by operating the shutter once, twice, three times, etc., with the bulb at room temperature. This gave us a record of the reflecting power of the inner surface of the quartz plate. The bulb was then raised to a red heat and another series of exposures made in the same way.

The plate showed that the rectangle representing the 2536 line had the same intensity for an exposure of five seconds for quartz reflection (bulb cold), and one second for mercury vapor reflection (bulb hot).

Since the reflecting power of a surface of fused quartz is roughly 5% in the ultra-violet, this experiment shows us that the reflecting power of a surface of dense mercury vapor for the light of the 2536 line is not far from 25%, nearly that of most metals in the same region of the spectrum.

The next point to determine was the density at which selective or metallic reflection commenced. Various methods were tried, the following being the one finally adopted.

It is of course desirable to repress as completely as possible the ordinary, or vitreous reflection of the quartz surface. This can be done by polarizing the incident beam with its electric vector horizontal and setting the prismatic face of the bulb at the polarizing angle. Under these conditions the intensity of the two beams reflected from the outer and inner surfaces is reduced nearly to zero, and a very slight increment in the reflecting power of the inner surface due to the mercury vapor becomes at once apparent provided the reflecting power is increased regardless of the direction of plane of polarization as proved to be the case.

The divergence of the two reflected beams was so great that both were not received by the quartz lens of the camera, consequently the one reflected from the outer surface was made nearly parallel to the other by reflection from a piece of platinized glass at nearly grazing incidence. Two small images of equal intensity thus recorded simultaneously on the photographic plate, one above the other. The upper, due to reflection from the outer surface of the prismatic plate, the lower representing the reflection from the inner surface. The temperature of the bulb was now gradually raised, and a number of exposures of equal duration made at different temperatures, the plate being moved slightly between exposures. Figure 9, Plate VII shows the result of the final experiment. The first three exposures, a, b, and c, corresponding to temperatures of 180°, 210°, and 235° showed both images of equal intensity, in other words, no increment of reflecting power resulted from the presence of mercury vapor up to a pressure of 50 mms. (at 235°).

Exposure d, taken at a temperature of 270°, showed the lower image considerably brighter than the upper one. The pressure in this case was about 120 mms., and since we can infer that the effect would be noticeable at a slightly lower pressure, we are safe in saying that the first appearance of specular reflection by the vapor takes place at a pressure not very far from 10 cms. Exposures e and f were made at temperatures of 300 (pressure 25 cms.) and a still higher temperature which was beyond the reach of the thermometer. On the print these have practically the same intensity, but on the original negative the density of f is certainly double that of e.

As subsequent experiments showed that the dense vapor reflects polarized light in much the same way as a film of metal, it was of some interest to see whether the reflecting power of the quartz surface in contact with the vapor passed through a minimum before beginning to increase with increasing vapor pressure. In the case of metallic deposits on glass, the reflecting power of the glass is considerably diminished by very thin layers of metal, when the reflection is from the glass side, in fact it is reduced nearly to zero if the thickness of the metallic film is just right.

If a number of strips are silvered cathodically on the face of a prism of small angle (five to ten degrees) with exposures to the discharge of from say one minute to ten minutes, one or more strips will be found which appear quite black in reflected light when viewed through the glass, though all reflect much more powerfully than the glass when viewed from the silver side.

The mercury vapor was examined for a similar phenomenon in the following way:

The convergent 2536 beam from the monochromator was reflected from the prismatic plate of the bulb and the two reflected beams received on a plate of uranium glass. The temperature of the bulb was then gradually raised and the fluorescent images on the uranium plate watched. If the vapor behaved exactly as a metal film of increasing thickness, the image formed by reflection from the inner surface ought to fade away gradually and then rapidly brighten. No trace of such a phenomenon was observed. The two images remained of the same intensity until a temperature of 250° was reached; above this point the image due to the inner reflection rapidly brightened as the temperature rose, reaching its maximum brilliancy in the neighborhood of 300°, at which temperature it appeared to be four or five times as bright as the other image.

This specular or metallic reflection of the light by the vapor occurs only when there is exact synchronism between the luminous vibration and the free period of the system which causes the 2536. This fact is emphasized because there is another type of selective reflection which occurs when the synchronism is not exact, and which is the result of the refractive index of the vapor. This will be discussed presently, after the polarization experiments have been treated.

POLARIZATION EXPERIMENTS

Renewed attempts have been made to detect traces of polarization in the scattered resonance radiation, but without success. Even when the vapor is illuminated with plane polarized light, and the spot of surface luminosity (which we have at say 100°) is photographed through a Savart plate or Fresnel double prism of R and L quartz with a polarizing analyzer of quartz and Iceland spar, no trace of the fringes appear. The same thing occurs in the case of sodium vapor illuminated with a sodium flame. This is very remarkable, since strong polarization has been found associated with the stimulation of the frequencies corresponding to the band or channelled spectra of sodium and iodine vapor, the polarization showing not only in the line directly excited by the monochromatic light (resonance radiation), but also in all of the other lines (lines of the resonance spectrum).[1]

In our present search for traces of polarization we employed the Fresnel double prism of right and left handed quartz. This gave, with polarized green mercury light, on analyzation, six dark horizontal bands. With the 2536 light the number of bands was found to be about thirty. They photographed very distinctly, however, with the quartz camera. As an analyzer we used a double image prism of quartz and Iceland spar, which we made by grinding and polishing a prism of about 8° from a small piece of spar, securing direct vision and fair achromatization by compensating it with a small quartz prism of about the same angle. Both prisms were cut, ground, and polished in less than an hour. The double prism was mounted in front of the quartz lens of the camera and the aperture considerably reduced by a diaphragm to secure sharp definition of the fringes.

In our preliminary experiments we employed a small Foucault prism as polarizer, but this reduced the intensity of the light enormously, and we accordingly cut a 60° prism of quartz perpendicular to the axis which, when mounted about 15 cms. behind the monochromator, in the converging beam of 2536 light, gave

[1] R. W. Wood, *Philosophical Magazine*, July, 1908, and October, 1911, p. 480.

two brilliant polarized images of the slit separated by a distance of about 3 mms. One of these was cut off by a screen, and the rays diverging from the other illuminated the prismatic plate of the bulb. The arrangement of the apparatus is shown in Figure 5, which explains itself. The camera is of course focused upon the Fresnel prism, by adjusting the distance as described previously.

An exposure of four or five minutes was sufficient to give the two images of the spot of resonance radiation formed by the double image prism.

At first we obtained distinct traces of the fringes even with the bulb at room temperature, while in some cases at higher

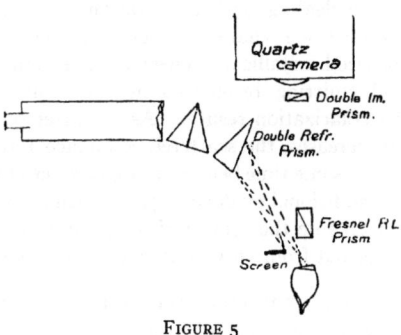

FIGURE 5

temperatures they were quite strong. Very variable results were obtained, however, and we finally found that they were invariably due to a slight mist of mercury globules deposited on the inner surface of the plate, or to light scattered by a slight cloudiness of the inner surface near one edge. We finally got things so adjusted that practically no trace of the fringes appeared, and feel quite certain that there is no trace of polarization in the scattered radiation, even when the incident light is polarized and the density of vapor is so great that we are approaching the stage at which metallic reflection begins. This seems very remarkable, since, as we shall see presently, if the incident light is polarized, the metallically reflected wave is polarized also.

This makes it again seem probable that we must reject the idea that the metallically reflected wave is obtained by applying the principle of Huygens to the waves emitted by the resonators. We had hoped to find that, as the vapor density increased, an increasing percentage of polarization would be found in the scattered radiation, but such does not seem to be the case.

It is, in fact, difficult to explain the absence of polarization. A rapid rotation of the molecule would probably act as a depolarizing factor. If the vibrations of the electron could take place only along a line, or was confined to a plane, and these lines or planes were oriented in all possible positions, we should expect less than 50% of polarization even with the incident light completely polarized. This was found to be the case with the resonance spectra of sodium and iodine. But, in this case, in which we appear to be dealing with a vibration of a much simpler type, we have no trace whatever of polarization. If we are to regard the polarized metallically reflected wave as the resultant of closely packed emitting resonators in synchronism, we must show how the polarization results. As a matter of fact, as the vapor density increases, the scattered resonance radiation practically disappears, some time before the appearance of the reflected wave, which is additional evidence against such a view.

We will now take up the subject of the polarization of the light metallically reflected from very dense mercury vapor.

POLARIZATION OF METALLICALLY REFLECTED LIGHT

In the case of the reflection of polarized light by metal surfaces, if the plane of polarization is parallel or perpendicular to the plane of incidence, the light remains plane polarized after reflection. If polarized in an azimuth of 45° the reflected light is usually more or less elliptically polarized.

The case has been examined in which the light is polarized with its electric vector at 45° to the plane of incidence. The light of the water cooled arc was passed through a small Foucault prism, arranged to transmit vibrations inclined at 45°. It was then reflected from the inner surface of the prismatic plate of the bulb, (which was set at the polarizing angle), into a quartz spectrograph. Between the slit and collimating lens, and close to the latter the small double image prism was mounted and oriented, so as to transmit horizontal and vertical vibrations, the images formed lying one above the other. In working with a quartz spectrograph it is important to analyze the polarized light before it enters the lenses, on account of the natural rotation of the latter.

The rays which entered the spectroscope were of course plane polarized by the reflection from the inner surface of the prismatic plate, or in other words, the incident polarized light was split into two components, one of which was wholly transmitted, while the other was in part reflected, with its vibrations perpendicular to the plane of reflection. This gave a single image in the analyzer. The slit of the spectrograph was opened somewhat, and its length contracted to such a degree that the spectrum lines appeared as small squares. Exposures were made with the bulb at room temperature. Figure 10, Plate VII (upper spectrum) and at a red heat (lower spectrum). In the latter case, owing to the metallic reflection of the 2536 line, both polarized vectors are reflected with equal facility, and unite into a plane polarized vibration at 45° azimuth, which is doubly refracted by the analyzer yielding two images of the square for the wave-length in question. It will be noticed that all of the other spectrum lines (squares) are represented by single images.

This experiment does not prove, however, that the reflected light is plane polarized, for the spectrum would have the same appearance if the reflected light was completely depolarized or circularly polarized. To prove that it is plane polarized we must rotate the double image prism through 45° and see if one of the images disappears. This was found to be the case. The prism was turned nearly to 45° and a number of successive exposures made, the prism being rotated through a small additional angle each time. One of these exposures showed the second image completely absent, proving that elliptical polarization was not present.

A very complete treatment of the 'Scattering and Regular Reflection of Light by Gas Molecules' has been given by C. V. Burton (*Philosophical Magazine*, May and June, 1915), in which the ratio of the scattered to the reflected radiation has been calculated. As I have already intimated, however, it seems doubtful whether we can apply these calculations in the present instance. Burton offers in explanation of the absence of polarization in the scattered radiation the hypothesis that much of it is of secondary or tertiary origin. The vapor has been excited with plane polarized light at nearly normal incidence, and the resonance radiation photographed in a nearly normal direction, not only at the very low pressure which it has at 0°, but also at various other pressures, up to that at which the light comes only from a very thin layer of the vapor in contact with the inner surface of the quartz plate. In no case has any trace of polarization been found.

This scattered and unpolarized radiation appears to vanish entirely (with further increase of density) and the reflected wave, which eventually appears, is completely plane polarized.

It does not appear to me that Burton's hypothesis is sufficient to explain the absence of polarization. With a resonator of Type II considered by Burton, which is identical with the type which I assumed in an earlier paper to explain the only partial polarization of sodium vapor, we should expect sufficient polarization to show the interference fringes, even if some secondary radiations were present.

If the *emission* of light results from a vibration set up in the molecule by the return of an expelled electron as assumed by Stark in certain cases, the absence of polarization is at once accounted for: the expulsion must, however, result from some sort of resonance vibration set up by the exciting rays, and unless we make the highly improbable assumption that this vibration gives rise to no emission of energy, it seems strange that there is no trace of polarization in the emitted light.

The presence of true absorption, which is neglected in Burton's treatment, of course complicates things very much. It undoubtedly increases at a rapid rate with increasing vapor density, and it may be present even at the density corresponding to room temperature.

It is, of course, very important to determine whether it is present at the lowest density which can be utilized.

We have tried the experiment, suggested by Burton, of detecting the heating effect, by means of a suspended vane, but failed to observe any effect, even when some air was present. The available energy appears to be too small.

The only method that has occurred to me involves a somewhat complicated calculation. In the absence of true absorption it should be possible to calculate, under specified conditions, the ratio of the secondary to the primary resonance radiation. A rough estimate of this ratio was made in my Guthrie lecture before the London Physical Society, and though it agreed fairly well with the observed ratio, it was little more than a guess. There is one other possible method also: the one described in the paper written in collaboration with Dunoyer, on the resonance of sodium-vapor. (This Monograph.) If a small patch of magnesium oxide on the surface of the resonance bulb is illuminated by the light of a resonance lamp it should not appear brighter than the resonating vapor, provided the scattering is complete (*i.e.*, no absorption). In the case of sodium vapor, illuminated by the light of a sodium resonance lamp, we found that the scattering power (diffuse reflecting power) of the vapor was almost as great as that of magnesium oxide.

In the case of mercury vapor, for the 2536 radiation it appears to be much less, the oxide coming out two or three times as bright as the vapor in the photograph. We used for the experiment a mercury resonance lamp of a type which will be presently described.

SELECTIVE REFLECTION AND REFRACTIVE INDEX

Selective reflection of another type occurs at the boundary surface separating quartz from dense mercury vapor. This occurs in the case of frequencies slightly higher than that of the 2536 line. The mercury resonators in this case emit no scattered radiation, and there is practically no loss by absorption.

In the earlier work, in studying the reflection of the light of the iron arc by mercury vapor, it was found that an iron line one Ångström unit on the short wave-length side of the 2536 mercury line was much more powerfully reflected than a pair of iron lines on the long wave-length side situated at 0.1 and 0.4 Å. U. from the mercury line. No explanation of this was given in the paper but the suggestion was made later in Wood's *Physical Optics* (second edition), page 432, that it undoubtedly resulted from the sudden change in the refractive index of the vapor in the vicinity of the absorption line.

"The 2536 line shows powerful selective dispersion and the refractive index, in its immediate vicinity on the short wave-length side, is much below unity, probably as low as 0.5 or even much less close to the line.

"In the case of light going from a rare to a dense medium, a high value of the refractive index for the latter is accompanied by strong reflection.

"When, however, the ray goes from dense to rare (quartz-mercury vapor) as in the present case, a low value of the index for the latter is accompanied by strong reflection.

"On the long wave-length side, for a region very close to the line the index of dense mercury vapor may rise to a value as high as that of quartz, in which case there will be no reflection at all."

It would appear then that, if we could employ light of two frequencies, one slightly higher and the other slightly lower than the frequency of the 2536 line, the former would be powerfully reflected and the latter not at all. This condition was realized by employing as our source of light a quartz mercury arc operated at a potential just sufficient to distinctly double the 2536 line by self-reversal.

In Figure 11 (Plate VII), we have four views of the 2536 line taken with a small Fuess quartz spectrograph, very accurately focused. This line has a faint companion on the short wavelength side, indicated by an arrow in the photographs. If the light is first passed through mercury vapor in a heated quartz tube, the main line is weakened or removed by absorption, and the faint companion remains as shown by (a) Figure 11, in which the upper and lower figures represent the line without and with mercury absorption. (b) Shows the appearance of the line, when the quartz arc, designed to operate at potential drop of 170 volts, is run at 30 volts, while (c) and (d) show it reversed at 60 and 80 volts.

We made our experiment as follows: The light of the lamp running at 80 volts was reflected from the inner surface of the prismatic plate of the quartz bulb into the quartz spectrograph, the slit of which was reduced to a length of 1 mm. by a diaphragm which could be raised by a micrometer screw. An exposure of one minute was given: the slit diaphragm was then raised 1 mm., the quartz bulb raised to a red heat by a Bunsen burner, and a second exposure of fifteen seconds made. Figure 6 (Plate VII) shows the result of this experiment. The reversed 2536 line appears as a doublet and is indicated by an arrow, the faint comparison on the short wave-length side appearing to its left. This was the exposure made by light reflected from the cold bulb.

Above it we have the exposure made with the hot bulb. The light reflected from the hot bulb is seen to consist solely of the short wave-length component of the doublet (widened and reversed 2536 line), for which the reflecting power of the quartz-mercury vapor surface is very high. The long wave-length component has disappeared entirely, owing to the very low value of the reflecting power for this frequency. The width of the doublet is about 0.8 Å.U. It is perhaps worthy of mention that we have here a rather efficient method of isolating from the total radiation of a quartz mercury arc running at a moderately high temperature, a single line of wave-length about 0.4 Å. U. less than that of the 2536 line of a similar lamp running at a low temperature.

This might be very useful in certain special investigations. We could of course make the difference even less than 0.4 Å.U. by operating the lamp at a lower voltage.

A powerful spark discharge between electrodes of cadmium was substituted for the mercury arc, and the light reflected from the inner surface of the prismatic plate of the bulb into the quartz spectrograph. Two spectrograms were taken in coincidence, one with the bulb cold, the other with it red-hot. The latter showed a bright line on the continuous background of the cadmium spark spectrum at wave-length 2536, bordered by a dark line on the long wave-length side. A photograph of a small portion of the spectrum of the light reflected from the hot bulb, in coincidence with the 2536 Hg. line is reproduced on Plate VII, Figure 12.

The bright, line of course, results from the powerful selective reflection by the vapor of a very narrow spectral range of the continuous spectrum: The dark border is due to the low reflecting power for the region-adjacent. Position indicated by A in figure.

We will now take up some points which were studied in the earlier investigation, considering first the amount of energy abstracted from the primary beam by the resonating gas molecules.

It is clear at the outset that if we wish to determine the amount of energy diverted by the resonators when they are in exact synchronism with the light-waves, it is useless to make observations upon the intensity of the light after it has suffered transmission through the vapor, even if we are dealing with what we are accustomed to call monochromatic light. All spectrum lines have a finite width, and the particular frequency scattered by the resonating molecules may constitute but a small fraction of the total energy of the spectrum line used to excite the vapor; in other words, it is only the center of the line that is effective in exciting resonance, the edges of the line not being reduced in intensity by the transmission through the gas. What we wish to determine is the reduction in intensity of that portion of the line, or in other words, the frequency, which is capable of exciting the natural period of vibration of the molecule. It appeared to me that the most direct way of investigating this question was to take the intensity of the cone of light as the measure of the intensity of the primary beam, for there appears to be no doubt but that the intensity of the resonance radiation is proportional to the intensity of that particular frequency in the exciting light which is capable of setting up resonance. It is obvious that we must use a beam of parallel rays in this case, and as it was desirable to be able to measure the density of the image of the resonating vapor close up to the inner surface of the plate, where the exciting beam entered, a brass box, furnished with two windows of crystalline quartz was employed in place of the tube with flared ends. This eliminated also a possible error that might have resulted from the circumstance that fused quartz is somewhat phosphorescent under the action of the ultra-violet rays. A very large number of photographs were made with different times of exposure and different vapor densities, and the photographic density of the image at different distances from the point where the light entered the vapor was measured. This gives us a measure of the rate at which the vapor cuts down the amplitude of the exciting frequency as the wave moves through the medium.

The results of these measurements showed that the intensity of the frequency capable of exciting resonance (the central part of the 2536 line) was reduced to about one-half of its value after traveling for a distance of 5 mm. in mercury vapor at a pressure of 0.001 mm.

This gives us a means of determining roughly the amount of energy diverted from the primary beam by each molecule, if we assume that all of them are equally effective.

Lamb, in his theoretical treatment of the absorption of light by a gas, published in the Stokes Commemoration of the Camb. Phil. Soc., sums up a calculation in the following words: "Hence in the case of exact synchronism, each molecule of gas would, if it acted independently, divert per unit of time nearly half as much energy as in the primary waves crosses a square whose side is equal to the wave-length." This means, if I am not mistaken, that if we had a density such that there was one molecule in each cube the sides of which were equal to the wave-length, the intensity of the light would be reduced by one half by traversing a single layer of molecules, while a density ten or twenty times as great as this ought to give selective reflection, since the wave would be practically stopped before penetrating to a depth of more than a small fraction of a wave-length.

Let us now compare this calculation with the values which have been determined. At a pressure of 0.001 mm., which is about the pressure used, the average molecular distance is such that we shall have on the average one molecule of mercury in every cube the sides of which are only very little larger than the wave-length (or, more exactly, 0.0003 mm.), which quantity divided into 5 mms., the distance traversed for a reduction of intensity equal to one-half, gives us 16,000, that is to say, 16,000 molecules must be passed before one-half of the energy is removed from a square element on the wave-front measuring λ on each side.

Of course this calculation is made on the assumption that all of the molecules are equally effective in scattering the light. It is, however, possible, even probable, that but a small percentage are, at any given moment, in the condition to act as resonators. Experiments on the dispersion and magnetic rotation of metallic vapors and luminescent hydrogen give evidence that but a small percentage of the molecules are at any instant concerned in the production of the phenomena in question.

PRIMARY AND SECONDARY RESONANCE RADIATION

Photographs of the luminous cone of mercury vapor at room temperature contained in the quartz tube appeared to prove that the vapor outside of the cone of vapor directly excited by the primary beam was itself luminous. It was observed, however, that the fused quartz phosphoresces with a violet light under the influence of the ultra-violet light, and I did not feel perfectly sure that the light did not come from the wall of the tube. To eliminate such a possibility a hollow box of brass was constructed (see Figure 6), two adjacent sides of which were left open, and closed with thin plates of quartz (crystal) which is not phosphorescent. The inside of the box was heavily smoked, and the plates cemented in place with sealing-wax. A drop of mercury was introduced and the interior of the box put in communication with a Gaede pump and exhausted. The ultra-violet light was focused at the center of the box, entering through one of the quartz plates, and the resonance radiation photographed from

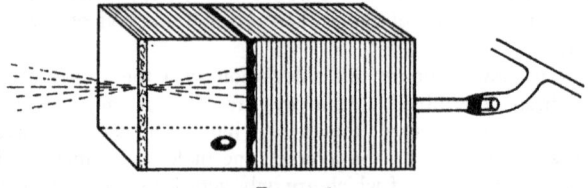

FIGURE 6
(From first Adams publication)

the side through the other plate. It was found that, unless the pressure of the air was less than three or four millimeters, no trace of any secondary radiation was present. On lowering the pressure of the residual air it developed rapidly, however, and after the pump had been on operation for several minutes, the luminous glow filled the entire interior of the box, the luminous cone being nearly lost in the strongly luminous background. With correctly timed exposures the directly excited resonance radiation is always four or five times brighter than the secondary radiation. Over exposure may, however, increase the effect of the secondary until it equals that of the primary, causing the outlines of the primary beam to disappear almost completely, as in the first picture in Figure 3, Plate VII.

The intensity of the secondary radiation depends upon the cross-section of the primary beam, as does also the rate at which its intensity diminishes with increasing distance from the primary rays. With an exciting beam of square cross-section (5 mm^2.)

the intensity of the secondary radiation half a millimeter from the edge of the beam was found to be nearly one-third of the intensity of the adjacent primary radiation. Its intensity fell off with increasing distance as follows:

Distance	Ratio
0.5 mm.	1/3
1.5 mm.	1/6
2.5 mm.	1/10
3.5 mm.	1/30

Four photographs of the phenomenon are reproduced on Plate VII, Figure 3. A vacuum-tube was put in circuit with the tube leading to the pump, to serve as an auxiliary manometer, and it was found that the secondary radiation did not appear at its maximum intensity until the green fluorescence due to cathode rays appeared in the vacuum-tube.

It is clear from the photographs that when the mercury vapor is in the highest possible vacuum, the light which it emits is capable of exciting a secondary radiation in the surrounding vapor which lies wholly outside of the path of the primary exciting beam. The presence of air at 4 or 5 mm. pressure, while it materially decreases the intensity of the primary resonance radiation, causes the secondary radiation to disappear entirely. In the four photographs which are reproduced I have recorded the pressure of the residual air in each case. To make sure that the disappearance of the secondary radiation was not due to a weakening of the primary radiation, I gave an exposure of four times the normal one, with air at 4 mm., and though the cone was very much blacker on the plate than on any of the others, there was no trace of any luminosity in the surrounding vapor.

This action of a small trace of air is most remarkable, and it is of the utmost importance to determine the explanation. Although the vapor which is in the path of the primary beam glows with almost undiminished brilliancy, the light which it gives out seems powerless to excite the surrounding vapor to luminosity.

The cause of this action of a chemically inert gas in destroying the secondary resonance radiation was finally found to be connected with the introduction of the factor of *true absorption*, as distinguished from molecular scattering of the light.

It appears that the introduction of air not only causes the disappearance of the secondary radiation coming from the regions lying outside of the cone of exciting rays, but also decreases the intensity of the radiations emitted by the mercury molecules lying in the path of the exciting rays.

In other words, as the pressure of the air, with which the mercury vapor is mixed, increases, more and more of the energy abstracted from the primary beam is *absorbed*, or converted into heat, and less and less is *scattered laterally*. The *total amount abstracted* is probably not much affected by the admission of air at these low pressures: if any change occurs it will be a slight increase, for it was found in an earlier investigation that the admission of air to an exhausted tube three meters in length, containing mercury, caused the appearance of the 2536 absorption line in the spectrum of the transmitted light.

It will now be shown that the introduction of this factor of true absorption is sufficient to account for the observed disappearance of the secondary radiation.

MOLECULAR SCATTERING AND TRUE ABSORPTION RATIO OF THE TWO QUANTITIES

It is easy to see that, if true absorption occurs as well as scattering the intensity of the secondary radiation in comparison to that of the primary will be greatly diminished. When the vapor is in a vacuum of less than .01 mm. it is possible that the energy diverted from the primary exciting beam is all scattered, and no true absorption occurs. We should of course find what appeared to be an absorption line in the spectrum of the transmitted light, and yet the molecules would not be absorbing energy but merely diverting it from the primary beam and sending it out in all directions. The molecules lying in the path of the beam will glow with a certain intensity, while those which lie outside of the path of the beam will be illuminated by the radiating molecules which are directly excited, and will in consequence emit a light of a lesser intensity. Suppose now that by the introduction of air at a pressure of 5 mm. the intensity of the light emitted by the directly excited molecules is reduced to one-third of its original value, the rest of the abstracted energy being absorbed. By means of a threefold increase in the intensity of the exciting light we can raise this intensity to its original value, so that the same amount of light is available for the excitation of the secondary radiation as before. The intensity of the secondary radiation excited under these circumstances will, however, be only one-third of its former value, since two-thirds of the energy received from the directly excited molecules is transformed into heat by the true absorption which has been introduced by the presence of the air. The intensity of the secondary resonance radiation in comparison with that of the primary will consequently be much less (one-third) than when the mercury vapor was in a high vacuum. This hypothesis was tested by experiment and practically proven. In the first place a very careful series of measurements was made of the reduction in the intensity of the primary resonance radiation by the introduction of air. The pressure of the air was measured with a McLeod gauge, and the duration and intensities of the excitation were made as nearly equal as possible.

The exposures were all made on the same plate, which was pushed along in the supporting clamps, the mercury lamp being allowed to cool down completely between exposures. To insure against accidental errors, a large number of plates were exposed, and the measurements made from each were compared. One of these plates is reproduced on Plate VII, Figure 1. The air pressure in the cell is marked on each picture. The exciting beam enters the cell from the right, and two-thirds of the quartz window was screened off, so that a number of exposures could be made on the same plate. In the first picture (pressure 0.01 mm.) the secondary radiation from the region not excited by the primary beam is very conspicuous, less so in the second and nearly gone in the third. The intensities of the primary radiation at the point where the incident beam entered the cell was measured by comparing the density of the negative with the density of a plate exposed in strips for times increasing gradually from 5 to 300, which plate was developed simultaneously with the other.

The values given are in the following table:

Air Pressure in mm.	Intensity of Primary Resonance Radiation	Absorbed Energy (assumed)
.01	300	0
.45	230	70
1.10	200	100
2.20	170	130
6.20	100	200
9.50	70	230
14.20	50	250
18.00	40	260
32.00	12	288

If we plot these values, taking intensities as ordinates and air pressures as abscissæ, we obtain a curve practically identical with the curve obtained with iodine vapor, which shows that the effect of the air upon the intensity of the emitted radiation is about the same in the two cases. In the third column I have given the amounts of the energy absorbed in each case. These values are merely the differences between the amounts of the emitted energies and the energy emitted when the vapor is in a high vacuum (300), and are calculated on the assumption that the total energy diverted from the primary beam is the same in the two cases, *i.e.*, that the presence of the air does not influence the amount of energy removed from the beam by the resonating gas molecules.

That this is in reality the case was shown by the following experiment. A double cell, Figure 7, was made by soldering across partition along a diameter of a short section of large brass tube, the ends of which were closed with quartz windows. The length of the tube was 17 mm., and the diameter 30 mm., and two small brass tubes permitted either compartment to be exhausted to any desired pressure. In measuring the energy diverted from the primary beam by the vapor, we must be certain that we use light which is in exact synchronism with the resonating molecules. The light must be far more homogeneous than the ray isolated by the quartz spectrograph from the light of the mercury arc. I used, therefore, what I shall hereafter refer to as the resonance lamp, a small quartz bulb, closed at the bottom with a flat plate of polished fused quartz, which was fused on in the same manner as the end plates of the long tube previously described. This bulb contained a drop of mercury, and was highly exhausted and sealed. The light from the quartz spectrograph was focused through the side of the bulb as close to the center of the flat bottom as possible. The adjustments were made by means of a small piece of uranium glass, which enables one to locate the path of the rays by its phosphorescence. It is most important to prevent the light which is reflected from the walls of the bulb from getting at the photographic plate. This gave a good

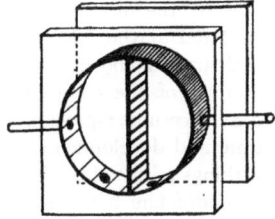

FIGURE 7
(From first Adams publication)

deal of trouble, but by means of the following device it was practically eliminated. A wooden box was made, measuring 40 x 10 x 10 cm., and a large circular hole cut in one end which was covered with a cone of black cardboard made by cutting out a circular disk, cutting along a radius and pasting the cut edges together, overlapping them about 2 cm. A small round hole was burned through the apex of the cone, and this was placed against the flat bottom of the resonance lamp. A quartz lens mounted in a partition of the box rendered parallel the rays which came from the resonance lamp through the small hole, so that the intensity of the light after its passage through the cell could be recorded close to the cell or at a distance from it. The importance of doing this is apparent from the following considerations. If we place a photographic plate close against the double cell containing the vapor, it will be illuminated by the primary beam which has traversed the cell and also by the scattered resonance radiation.

If, however, we place the plate at a distance, say at the other end of the box, the primary rays, being parallel, will reach it with undiminished intensity, while the effect of the scattered radiation will be negligible, since its intensity diminishes according to the law of inverse squares. No difference should be found with the plate in the two positions for the light which has gone through the compartment containing air and mercury vapor, since, as we have seen, the presence of the air destroys the resonance radiation. The experiment was made in the following way. One compartment of the double cell was highly exhausted and the other to a pressure of 3 cm. A strip of photographic plate 1.5 cm. in width was mounted close to the cell and received the light which had traversed the lower half of each compartment. A larger plate was mounted at the other end of the box, and received the light which passed above the first plate, and had traversed the upper halves of the compartments. Thus four records were obtained at once under precisely similar conditions as regards exposure-time and development. Much trouble was experienced in getting things adjusted so that the intensity close to the lens and at the endof the box came out the same with the cell removed, which is of course a necessary preliminary experiment. It was finally found that the air of the room contained enough mercury vapor to reduce the intensity of the light from the resonance lamp by nearly one-half as it traversed the length of the box.

This is not so surprising when we remember that the earlier experiments showed that the primary beam was reduced to one-half of its intensity (*i.e.*, the frequency capable of exciting resonance was) by traversing 5 mm. of the saturated vapor at room temperature. The trouble was overcome by opening the windows and thoroughly ventilating the room before each experiment. One is reminded of the trouble experienced in carrying on certain investigations in laboratories which have become infected by radium!

The photographs showed that the energy diverted from the primary beam was the same for mercury in a high vacuum and in air at 3 cm. pressure, so that the calculation of the absorption-scattering ratio which I gave provisionally was justified.

It was found, however, that if air at atmospheric pressure was admitted to one compartment the absorption was decreased by more than one-half, which shows that the air, in addition to broadening the absorption line, reduces the intensity of the absorption at its center. I have observed the same thing with iodine vapor, the lines becoming fuzzy and less black when air is admitted to the tube.

EXPERIMENTS WITH THE RESONANCE LAMP

The radiation emitted from the exhausted quartz bulb containing a drop of mercury is so homogeneous, that a layer of mercury vapor 5 mm. thick and at the pressure which it has at room temperature (0.001 mm.) reduces its intensity by about one-half. Various investigations with the vapor at exceedingly low pressure at once became possible. It is as if we had a gas which appeared quite black even at pressures commonly employed in vacuum-tubes. It will be possible to study the rate at which the vapor diffuses into other gases at low pressures, and it may be possible to tell in this way whether the resonators are in reality mercury molecules or larger aggregates.

I made two photographs which illustrate what a sensitive detector of small traces of mercury vapor we have in the light of the resonance lamp. A quartz bulb having an internal diameter of 1.5 cm., containing a drop of mercury, was mounted in front of a photographic plate in a dark box and illuminated with the light of the lamp. The bulb cast a shadow as black as ink. The bulb was then carefully freed from every trace of mercury and again photographed. The two photographs are reproduced on Plate VII, Figure 4, one with the flask empty, the other when filled with mercury vapor at room temperature.

I next drilled a shallow cavity in the end of a brass cylinder, warmed it to a temperature of perhaps ten degrees above the temperature of the room, and placed a drop of mercury in the cavity, the drop standing up above the level of the end of the cylinder. This was photographed in the dark box by the light of the resonance lamp, and the picture showed the black column of mercury vapor carried up by the convection current of warm air.

As a resonance lamp we have more recently used the quartz tube, closed by worked plates of fused quartz (previously described), containing a drop of mercury and highly exhausted. This tube was mounted in front of, and close to, the crystalline quartz plate which formed the window of the water-jacket of the lamp. It is important to have the rays of the arc traverse the mercury vapor as near to the front window of the resonance lamp as possible, since it has been shown in one of the previous papers, that the intensity of the resonance radiation is reduced to one-half of its value by traversing a layer of mercury vapor at room temperature, only five millimeters in thickness. A screen of black paper, perforated with a hole cuts off stray radiation scattered by the walls of the resonance lamp, and it is advantageous to cover the further end of the tube with a small cap of black paper, or provide some other suitable black background.

If the invisible light from the resonance lamp is focused upon a sheet of uranium glass by means of a large quartz lens we obtain a bright spot of yellow fluorescent light and can render visible the vapor rising from a warm drop of mercury by holding it close to the screen in the path of the rays, the shadow of the vapor cast on the uranium glass, appearing like a column of black smoke, as shown in one of the photographs published in the earlier paper.

Still better than uranium glass is a screen of barium platinocyanide. The phosphorescence of this substance varies according to the manner of crystallization. I have obtained good results by forming a saturated solution in water at 50° containing a little barium cyanide, immersing the beaker in ice water and stirring the solution vigorously with a good sized paint brush. This causes the formation of very minute crystals.

No. 12

Separation of Close Spectrum Lines for Monochromatic Illumination

In many branches of research in physical optics it often becomes necessary, for one reason or another, to separate two or more close spectrum lines, utilizing the light of one only.

For example, in experiments upon the monochromatic excitation of resonance spectra, the line utilized for the illumination of the fluorescing vapor must be isolated either by absorbing screens or by a spectroscope, used as a monochromator.

If the latter method is employed the illumination is much restricted by the necessity of employing a slit, or rather two slits; and in the case of close spectrum lines, such as the D lines of sodium, the necessity of employing very fine slits makes it very nearly impossible to accomplish anything in this way. Even in the case of the three green copper lines, I found the greatest difficulty in getting sufficient illumination with a single line isolated by means of a very large monochromator of 1.5 meters focus.

In the present paper I shall give a method which enables us to utilize a source of light of large size, say 1 × 3 cm., and remove one or more lines from it with *practically no loss of light*.

For example, we can form three images of a sodium flame by means of a condenser having an effective aperture equal to $f2$, one image containing only the light of wave-length 5890, the other two only light of wave-length 5896, the former being very nearly as intense (with respect to *one* sodium line) as if the condenser had been employed without the separating apparatus.

The method is an improvement upon one which I used many years ago in the study of the dispersion of sodium vapor and described briefly at that time. It is a polarization method, and may be described briefly as follows.

If plane-polarized monochromatic light is passed through a plate of some doubly refracting crystal with its direction of vibration making an angle of 45° with the axis, it will emerge plane-polarized parallel to the original plane for certain thicknesses of the plate, and plane-polarized at a right angle to this plane for other thicknesses. For intermediate thicknesses it will be elliptically or circularly polarized.

If we employ a plate of quartz 30 mm. thick the emergent waves of D_1 and D_2 of sodium will be plane-polarized at right angles to each other, and either can be quenched by a nicol suitably oriented. If white light is used, and analyzed by a spectroscope the spectrum will be furrowed by dark bands, the distance between a bright and a dark band being, in the yellow region, 6 Ångström units, the distance between the D lines.

As it was desired to utilize this principle for the separation of the D lines for the purpose of exciting the resonance radiation of sodium vapor by the light of D_1 and D_2 separately, by which means we may determine whether the mechanisms which give rise to the radiations are coupled together, an investigation which has been carried to a successful conclusion in collaboration with L. Dunoyer and is described in another paper, it became necessary to bring the method up to the highest possible efficiency. As it is necessary to employ a large condenser and work with very divergent and convergent cones of light, a block of quartz of very large size must be used, placed between the two halves of the condenser, since the rays which traverse the block must be parallel. If this is not the case, different thicknesses of quartz will be traversed, and the emergent rays will be polarized in various azimuths. Moreover, one-half of the light is lost at the start by the polarizing nicol. This difficulty was overcome by employing a large double-image prism, and subsequently analyzing by a double-image prism. In this way, with proper orientation of the prisms, the two images containing only D_2 light were superposed, the D_1 images (of one-half the intensity) lying to the right and left. By this expedient the D_2 image had the full intensity, except for the loss by reflexion from the six transparent surfaces of the prisms and quartz block.

A rotation of 90° of the plane of polarization is produced by a quartz plate .032 mm. in thickness for sodium light consequently the plate must be plane-parallel to within considerably less than this distance, otherwise D_1 will be passed by some parts of the field and D_2 by others. If the difference in thickness changes by .032 mm. in passing from one edge of the block to the other, one edge of the field will transmit D_2 only, and will appear brighter than the other edge which transmits only D_1, while the center of the plate will transmit both D_1 and D_2 in a state of circular polarization. Here both wave-lengths will be passed by the analyzing nicol, and the intensity will be intermediate between the values at the edges. If the thickness varies at a more rapid rate, bright and less bright bands will cross the field, the bright bands representing D_2, the less bright D_1 light.

With monochromatic light, *e.g.*, the green light of the mercury arc, the bands are black. If the intensities of the D lines were equal, the bands would be invisible. They are more distinct with a feeble sodium flame than with one of great intensity, since for a feeble flame the intensity ratio $\frac{D_2}{D_1} = 2$, while for an intense flame $\frac{D_2}{D_1} = 1.3$.

These bands are also visible with a thick block of uniform thickness owing to the fact that the rays entering the eye from an extended source of light traverse slightly different distances in the quartz.

The calculated value .032 mm. was verified with a fragment of a quartz plate 30 mm. in thickness and slightly wedge-shaped, placed at my disposal by Mr. Twyman, manager of the firm of Adam Hilger & Co., who also loaned me the quartz echelon used in the preliminary investigation. With this block between crossed nicols seven bands were counted (counting both dark and bright). This means that we pass from D_1 to D_2 transmission seven times in crossing the plate. Multiplying the calculated thickness .032 mm. by seven gives us 0.224, while the difference in thickness of the plate at the two edges, as measured with the spherometer, was found to be 0.243.

An investigation was also made with an echelon of quartz placed with its elements horizontal between the nicols. The polarizing nicol is of course placed with one of its *diagonals* making an angle of 45° with the optic axis of the quartz. It was illuminated with a sodium flame, and an image of the steps thrown upon the slit of a spectroscope. Each element of the slit covered by an echelon step is thus illuminated by light which has traversed a different thickness of quartz. At some points only D_1 appeared at others only D_2, while at others both D_1 and D_2 were found.

A photograph of this phenomenon is reproduced on Plate VIII Figure 1. The best separation of D_1 and D_2 was given by step number 6, and as each plate was 4.7 mm. in thickness, the total thickness was 4.7×6 or 28.2 mm. The seventh step showed both D_1 and D_2, as the total thickness here happened to be that giving circular polarization for both wave-lengths. A slight inclination of a plate of this thickness would cause it to transmit D_1 or D_2 only, by changing the length of the optical path in quartz.

In practice we may use a plate anywhere between 25 and 40 mm. in thickness. The best thickness is 32 mm., which gives us the maximum intensity for either sodium line when the other is cut off. With a plate of say 25 mm. in thickness D_1 can be completely extinguished, but the transmitted light (D_2) will not be as bright as when a plate of the correct thickness is used.

If the echelon is illuminated with white light, the continuous spectrum transmitted by each step is furrowed by black bands, which represent wave-lengths of light vibrating parallel to the long diagonal of the analyzing nicol. The distance between the bands decreases with the number of plates which operate at each step. A photograph of these bands with the D lines superposed is reproduced on Plate VIII, Figure 4.

For a thickness equal to 32 mm. the distance between adjacent bright and black bands is 6 Ångström units. If a different thickness is employed, and a black band brought into coincidence with D_1, D_2 will lie a little to one side of the center of the adjacent bright band and its intensity will be less than if the correct thickness is employed.

The extinguishing of one of the D lines can be shown with a natural uncut crystal of quartz, if the surfaces are fairly good. The crystal is to be placed between crossed nicols, utilizing two opposed surfaces, which are separated by a distance of two or three centimeters.

An image of the crystal is projected upon the slit of the spectroscope, and D_1 will be found absent at certain points, D_2 absent at others. It is another matter if a large amount of light is to be used, as in experiments upon fluorescence, for in this case we must use a large block of uniform thickness *free from all traces of crystalline irregularities*. Brazilian quartz should be used, as the crystals from Madagascar show irregularities when examined by polarized light.

A large and very clear crystal was selected and examined between crossed nicols with a sodium flame diffused by a sheet of ground glass. There appeared to be no internal irregularities of a nature such as were exhibited by a beautiful block of quartz loaned to me by Mr. Twyman, which was possibly cut from a Madagascar crystal. In this block the D_1 and D_2 bands, instead of appearing uniformly parallel, were deformed at one point in sharp zig-zags; no trace of anything could be seen by unpolarized light. It is not always possible to judge a crystal before it is cut, but if the faces are reasonably plane and clear, it is usually possible to tell whether variations in the intensity of the sodium light results from internal troubles, or from small differences in thickness. From the selected crystal a block measuring 85 mm. $\times$ 60 mm. $\times$ 32 mm. was cut parallel to the axis and polished. The most sensitive test for optically perfect quartz is examination with a sodium flame between crossed nicols in a direction parallel to the optic axis. The majority of crystals show zig-zag bands or flakes due to crystalline irregularities. Perfect crystals show uniform illumination or circular arc according to the thickness examined.

As the degree to which the plate needs to be plane-parallel can easily be attained by the use of the spherometer, and as the side faces do not have to be accurately parallel to the axis, the preparation of the plate presents no great difficulty.

It was examined with sodium light between crossed nicols before the polishing stage was reached, and found to give no abnormalities of the D_1 and D_2 bands, which is what is required. Glass plates and benzine were of course used to render the block transparent. The uniformity of thickness was considerably increased even after this test, after which the six surfaces of

the block were polished. This block when used for separating the
D lines (32 mm. thickness) has an area of 50 square cms. Used in
the other position, 60 mm. thickness, it will separate lines 3 Ångström units apart. Used 'end-on' its property of natural rotation can be utilized in certain experiments.

The double refraction of Iceland spar is very much greater
than that of quartz, a plate less than 3 mm. thick being required
for the D lines. It would, however, have to be plane-parallel
to a degree 10 times as great as is the case with quartz. A thick
block of spar, made plane-parallel by the methods in use in the
construction of modern interferometers, could very likely be used
for suppressing the strong central components of multiple lines,

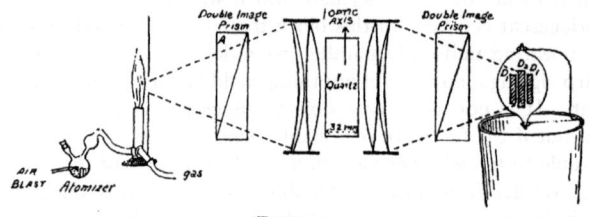

FIGURE 1

when photographing or examining the fainter satellites. Experiments in this direction are now in progress.

The quartz plate has proved most satisfactory in every
respect. Placed between the two halves of a Dunoyer condenser
of 10 cm. diameter and 12 cm. focus for parallel light (*i.e.*, 24 cm.
focal length when forming an image of a source of light equal in
size to that of the source), it is possible to utilize the entire 50
square cm. of the plate in forming an image of the aperture (1×3
cm.) placed before the sodium flame, containing only D_1 or D_2 light.

The arrangement of the apparatus is shown in the diagram.
The double-image prism A is so oriented as to have the vibrations of the two transmitted beams at 45° with the vertical.
The second double-image prism is oriented *in the absence of the
quartz block*, so as to give two images of the source side by side,
and separated by a distance equal to the width of each image.
Each image contains both D_1 and D_2. If now the quartz block is
placed between the two halves of the condenser, all of the D_2 light

(for example) leaves the original images, and unites into a third image between the other two. This is the image utilized. If it is desired to have D_1 light in this image, it is necessary only to rotate the block a degree or two, thereby increasing the optical path in quartz. If we rotate the analyzing double-image prism the central image splits up into two, and the correct position of the prism can be determined by bringing these two images into exact coincidence. As I have said previously, this method obviates the 50 per cent. loss of light which results from the use of Nicol's prisms. As the double-image prisms had apertures considerably less than that of the quartz block, they were placed midway between the condenser lenses and the image and source, as shown in the figure. With prisms of too small aperture, however, placed close to the source and image, the separation of the images would be insufficient.

To test the adjustment of the quartz block and the general efficiency of the apparatus, we have only to receive the three images upon a piece of white paper, and project an image of the central one upon the slit of a spectroscope capable of clearly resolving the D lines.

If the optic axis is not vertical some parts of the slit will be illuminated with D_1, others with D_2, and others with both D_1 and D_2, as shown by the photograph reproduced on Plate VII, Figure 2.

The completeness of the extinction is seen from the very faint trace of D_1, seen at the regions where D_2 only was transmitted. A very fine slit was used, and the plate overexposed.

The block must be tilted forwards or back until the same condition obtains all along the slit. The block is now rotated slightly until either D_1 or D_2 is completely quenched. We can now be sure that the image is made up entirely of monochromatic light, as shown in Figure 3, in which the upper portion of the slit is illuminated by D_2 light, and the lower by the light of the sodium flame.

For work with the spectroscope alone, in which cones of light of large aperture are not needed, a quartz echelon answers every purpose, so that this instrument, recently placed on the market by the Hilger Co., may be utilized in a new way.

The experiments outlined in the present paper were carried on at the Sorbonne, in the laboratory of Monsieur Bouty, who placed every facility at my disposal.

NOTE ON THE PRODUCTION OF A VERY INTENSE SODIUM FLAME

In the course of the experiments described in the preceding paper, it was found desirable to have an exceedingly intense sodium flame available for the adjustment of the quartz block for the complete extinction of one of the D lines.

As a flame of this description is often desirable in many branches of optical work, it has seemed worth while to add a separate note on the subject.

The intensity of a soda flame depends chiefly upon the rate at which the sodium molecules are delivered into the flame, that is the rate at which the chloride of sodium is volatilized. If a small fragment of the mantle of a Welsbach light is laid upon the grill of a Meker burner, and two or three small fragments of fused sodium chloride are placed on this, on lighting the burner a flame of a most astonishing brilliancy is at once formed. So rapid is the evaporation of the chloride that clouds of smoke rise from the flame, and the intensity, while at its maximum, appears to be as great as that of the oxy-hydrogen sodium flame, which is much more difficult to manage. The simplicity of this method makes it immediately available in any laboratory. The function of the scrap of mantle is of course to spread the material over a large surface of very small heat capacity, so that it can be brought to the temperature of the hottest part of the flame. The bead melts and the mantle acts like the wick of a lamp.

No. 13.

Photometric Investigation of the Superficial Resonance of Sodium Vapor

(In collaboration with L. Dunoyer)

The vapor of sodium, relatively cold, is capable of re-emitting the D line radiations, when one concentrates on it light containing these same radiations. This was demonstrated by one of us [1] in 1905, the image of an oxy-hydrogen sodium flame being formed by a large condenser along the axis of a highly exhausted tube containing a fragment of metallic sodium heated by a small Bunsen flame. At the same time it was shown that the cone of luminosity formed by the exciting rays retreated towards the wall, as the density of the vapor increased until there remained only a thin skin of yellow light, which lined the inner wall of the tube.

The experiment, as carried out at this time, was of short duration, and it did not appear possible to carry on any extensive investigations, with the disposition of the apparatus then employed.

The method of exciting the resonance has, however, recently been greatly improved by one of us [2] by using small glass bulbs, the walls of which are very carefully freed from occluded gases by prolonged heating, pure sodium being introduced into them by distillation. A further improvement consists in the employment of a Meker burner fed by the spray of a very dilute solution of sodium chloride as a source of light, and forming an image of it on the wall of the bulb by an aplanatic condenser of very large aperture. The sharpness of this image permits of a study of the phenomenon of *secondary resonance* discovered by one of us in the case of the vapor of mercury and described recently.[3]

[1] Wood, *Philosophical Magazine*, x., p. 513 (1905).
[2] Dunoyer, *Journal de Physique*, iv., p. 17 (1914).
[3] Wood, *Philosophical Magazine*, May, 1912.

The surface of the bulb, illuminated in the manner described, becomes the source of a resonance radiation of remarkable brilliancy, of a thickness too small to be observed; as the duration of the phenomenon is ten or fifteen hours, it may be investigated photometrically or spectroscopically without difficulty. The preliminary study showed that the intensity of the resonance is much greater if a flame relatively poor in sodium is employed, than with a powerful flame such as is obtained if a fragment of salt is placed on the grill of a Meker burner. If the bulb is heated by a large flame colored only by the sodium in the air of the room (previously charged by the operation of an intense soda-flame for a few minutes), one observes the yellow glow of the superficial resonance excited by the light emitted by the flame used for the heating of the bulb. The flame must be waved about rapidly over the surface of the bulb, in order to secure a fairly uniform temperature. The phenomenon is less marked if an intense sodium flame is employed.

These experiments show that the greater part of the D light of the flame is inoperative in exciting the resonance. Moreover, the intensity of the source appears scarcely diminished if it is viewed through the bulb in which resonance is taking place. In other words, it is only the central cores of the D lines which are effective in exciting the resonance. The same phenomenon was observed in the study of the resonance of mercury vapor already alluded to, the luminosity (ultra-violet) excited by the 2536 line being enormously greater when the exciting mercury arc was first started, than after it had been in operation for a few seconds.

In the present communication we shall discuss:

(1) The photometric study of the diffusing power of the highly attenuated vapor for monochromatic light, as compared with that of a white matt surface of magnesium oxide.

(2) The conditions under which all of the light removed from the exciting beam is re-emitted, giving us a diffuse reflecting power equal to that of the magnesium oxide.

(3) The probable width of the spectrum lines emitted by the resonating vapor.

APPARATUS EMPLOYED

The source of light for exciting the resonance was a Meker burner surrounded by an iron chimney perforated with a rectangular aperture. The burner was fed at the base with a spray formed by an atomizer operated by compressed air. A nearly saturated solution (30 grs. to the litre) of NaCl was diluted to $\frac{1}{32}, \frac{1}{64}, \frac{1}{128}, \frac{1}{256}, \frac{1}{512}, \frac{1}{1024}$ and $\frac{1}{2048}$, and these solutions introduced in turn into the bulb of the atomizer, previously well rinsed out with a solution of the concentration employed.

FIGURE 1

An image of the window in the iron chimney was formed on the wall of the bulb by an aplanatic condenser of the type described recently by one of us,[4] having a diameter of 11 centimeters and a focus of 12 cm. for parallel light. For divergent light, as in the present case, the source and image are each 25 cm. from the lens.

As the sodium bulbs used in these experiments are very easily made, and are extremely convenient for illustrating resonance radiation, it may be well to devote a few words to the manner of preparing them. The bulbs are 5 cm. in diameter, drawn down to 1 mm. capillaries as shown in Figure 1. The sodium must first be heated to fusion in a small test-tube and poured out on a cool surface. A piece about $2 \times 2 \times 2$ mm. is introduced into a small piece of very thin-walled glass tubing, closed at one end, and this capsule placed in the lower tube A, which serves as the distillation chamber. The lower end of A is now closed in the flame of a blast-lamp, and the tube sealed to the mercury pump. After exhausting to a pressure of 0.001 mm. the bulb is heated for four or five minutes with a large Bunsen flame, the pump working all the while. It should be heated as hot as possible without collapsing. After the bulb has cooled off, the flame is carefully applied to the chamber A, and the sodium distilled into the bulb. The lower capillary is then sealed, and finally the upper. The pump should be working vigor-

[4] Dunoyer, *Journal de Physique*, iii, p. 468 (1913).

ously all the while, as the brilliancy of the resonance depends upon having the highest possible vacuum. In our experiments we heated the bulb for twenty minutes, to make sure of getting rid of all of the gases, and the sodium was previously heated *in vacuo*, but these extreme precautions are not necessary in the preparation of bulbs for lecture purposes.

The bulb was supported by a wire in a column of hot air rising from a large tube of fireclay with a large Meker burner at the

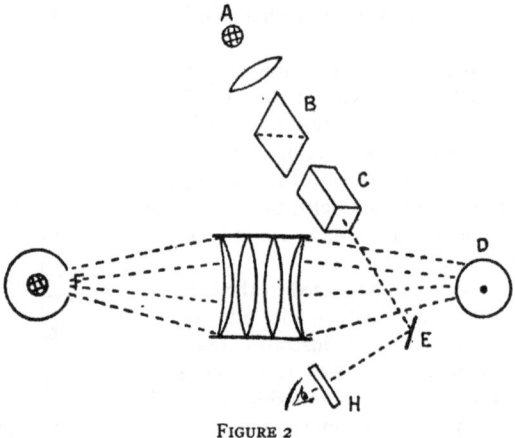

FIGURE 2

bottom, by means of which a fairly uniform temperature up to 400° could be obtained. For lecture purposes it is sufficient to heat the bulb by waving a large Bunsen flame over its entire surface. The arrangement of our apparatus is shown in Figure 2. As a standard source of comparison we used a sodium flame A behind a piece of ground glass mounted behind a pair of large Nicol prisms B and C. The sodium bulb D was first coated with magnesium oxide, by holding it above a piece of burning Mg ribbon. It was then wiped clean, a small square patch of the oxide being left on one side. The image of the window in the iron chimney surrounding the exciting sodium flame F was formed upon the square of magnesia and the adjacent clear glass of the bulb. It was our object to measure the ratio of the intensity of the magnesia and the vapor of sodium under equal illumination. This was done by means of a very simple photometer which consisted of a thin scale of silvered glass E, with a razor edge, made by silvering a piece of plate glass, polishing it, and then striking the edge with a hammer. This mirror reflected the comparison source A to the eye through a cell H containing a solution of bichromate of potash (to remove the green and blue rays of the Bunsen flame). Behind the sharp edge of the silver mirror the illuminated surfaces of magnesia and sodium vapor could be seen at the same time, and by adjusting the Nicol prism C the edge of the mirror could be made to disappear, first when seen against the magnesia and secondly against the background of resonating vapor. The intensities of the two surfaces are then in the ratio of the squares of the angles through which C is turned from the position of extinction. The temperature to which the bulb was heated by the ascending current of hot air was about 330°, measured with a nitrogen mercury thermometer.

The results are given in the following table, the concentrations of the salt solution in the atomizer bulb in the first column, the angles of the Nicol prism C in the next two columns, and the ratio in the fourth.

Solution Concentration	Angle a of Nicol for Resonance Radiation	Angle a' of Nicol for Magnesia	Ratio $\frac{\sin^2 a'}{\sin^2 a}$ $\frac{I' \text{ Magnesia}}{I \text{ Sodium vapor}}$
$\frac{1}{2048}$	3°	6°	4
$\frac{1}{1024}$	9.25°	19°	4.5
$\frac{1}{512}$	10.5°	22.5°	4.8
$\frac{1}{256}$	13.5°	36.4°	6.3
$\frac{1}{128}$	14.4°	45°	9.6
$\frac{1}{64}$	14°	67°	15
$\frac{1}{32}$	12.6°	90°	19

We see from this table that even with the most dilute solution the diffuse reflecting power of the magnesia is four times as great as that of the resonating sodium vapor, for the total yellow light of the flame. This is of course due to the circumstance that the magnesia reflects all of the D light, while the vapor scatters only the light corresponding to the cores of lines, the light of the edges of the lines being transmitted. As the concentration increases the intensity of the resonance radiation increases but slightly after a certain point is reached, since the gain in the intensity of the sodium flame then results chiefly from a widening of the lines. For the most concentrated solution ($\frac{1}{32}$) the magnesia was 19 times as intense as the vapor. On reducing the air-current until the yellow color of the flame was barely visible, a ratio of 3 was obtained, the values of the angles being 2° and 3°.5. This result was, however, open to question on account of the faintness of the light.

The above results are in accord with those previously obtained by one of us by a different method.[5]

If now the molecular resonators absorb none of the light which they abstract from the exciting beam, we ought, if the exciting radiations are made sufficiently homogeneous, to have all of the light diffusely reflected by the vapor; in other words, our ratio ought to sink to unity when the D lines in the source become infinitely narrow. It is impossible to reach this point by diminishing the amount of sodium in the flame, our lowest value for the ratio being four, or perhaps three.

We have, however, investigated the matter by employing the principle of the resonance lamp previously described by one of us, which has been used in the investigations on the resonance of mercury vapor. The experiment was made by utilizing the spot of superficial resonance as a source of light for exciting the vapor at a different point on the surface of the bulb. The arrangement of the apparatus is shown in Fig. 3. A small triangular spot of magnesia was formed on the surface of the sodium bulb with a black dot of lamp-black at its center to indicate its position. The image of the sodium flame was thrown upon this spot, and the magnesia triangle shone brilliantly upon the less intense background of the resonance radiation (Fig. 3, a).

[5] L. Dunoyer, *Journal de Physique*, iv, p. 17 (1914.)

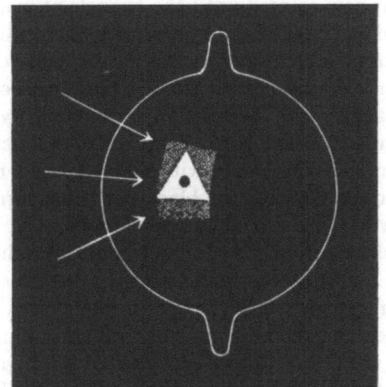

FIGURE 3
(a)

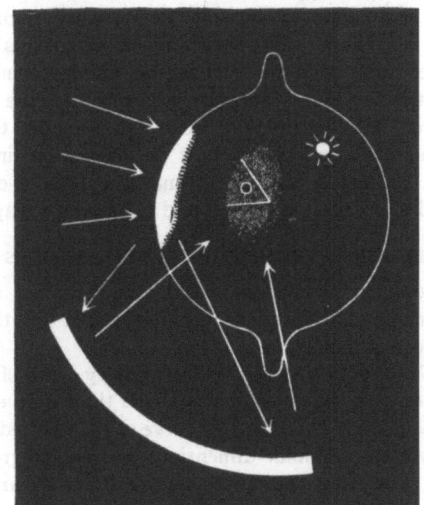

FIGURE 3
(b)

The bulb was now rotated until the triangle of magnesia was in darkness, and an image of the spot of resonance radiation thrown upon it by means of a large concave mirror, formed by silvering one surface of a double convex lens. Under these conditions the magnesia triangle, and the resonance radiation (which may be termed secondary) which surrounded it had practically the same intensity. In fact it was only with difficulty that the outline of the triangle could be seen, the black dot being surrounded by a uniform glow of light of oval outline (Figure 3, b).

Photographs taken of the phenomena are reproduced on Plate IX, Figures 8 and 9, the latter showing the disappearance of the magnesia triangle, when the area is illuminated by resonance light reflected from the mirror. The brilliantly illuminated area to the left is the primary resonance excited by the rays from the flame. A narrow dark line partially outlines the triangle; this is due to the shadow thrown upon the resonating vapor by the edges of the layer of magnesia.

The complete disappearance of the triangle was observed only when the flame for exciting the primary resonance was very poor in sodium. We thus have a ratio equal to unity when the exciting rays are sufficiently homogeneous, and can safely say that no true absorption exists in the case of sodium vapor at very low density and in a high vacuum, though the spectroscope would of course show an absorption line. All of the energy abstracted from the primary beam is re-emitted by the molecules, precisely as was found for mercury vapor.

PROBABLE WIDTH OF THE RESONANCE LINES

The experiments which we have just described show that the resonance radiation of sodium is excited by the narrow central regions of the D lines.

Let ABC of Figure 4 represent the intensity curve of one of the exciting lines, and the dotted curve DBE the region effective in exciting resonance. The intensity curve of the emitted resonance radiation will be of similar dimensions and may be represented by F. The ratio of the area of the curve ABC to the area of DBE is obviously the ratio found by the photometric measurements, and if we know the form of the curves, and the actual dimensions of ABC (i.e., the width of the line in the flame spectrum), we can,

Superficial Resonance of Sodium Vapor

from our experimentally found ratio of 4 : 1, determine the width of *DBE*, the line of the resonance radiation.

The interferential measurements of Fabry and Buisson have shown that the widths of the *D* lines emitted by a flame poor in sodium are 0.08 ÅU.

The law of the partition of energy in spectrum lines furnished by the kinetic theory of gases is

$$y = Ce^{-kx^2}, \quad \ldots \quad (1)$$

in which y is the intensity at a distance equal to x from the

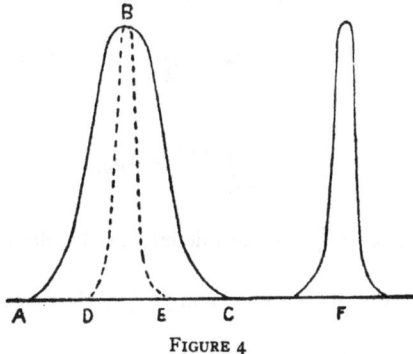

FIGURE 4

centre of the line (at which point the intensity is C), and k gives us the rate at which the intensity falls off as we depart from the center.

The width of 2ϵ of a spectrum line, as defined by Fabry and Buisson, is the distance between two ordinates at distances ϵ from the center, of value equal to $\dfrac{C}{2}$, *i.e.*, an intensity one-half as great as that at the center of the line.

This gives us

$$\frac{1}{2} = \frac{Ce^{-k\epsilon^2}}{C} \text{ and } e^{-k\epsilon^2} = 2, \text{ or}$$

$$\sqrt{k}\,\epsilon = \sqrt{\log 2}. \quad \ldots \quad (2)$$

The total intensity of the line being the area comprised between the curve and the axis of abscissæ, we have

$$I = \int_0^\infty Ce^{-kx^2}dx = \frac{C\sqrt{\pi}}{\sqrt{k}} \quad \ldots \quad (3)$$

If we assume that the portion of the exciting line effective in exciting the resonance is comprised between the ordinates situated at $-x_1$ and $+x_1$ (which is of course an approximation since in reality the exciting portion of the line is as shown by the dotted curve of Fig. 4), we have, for the intensity removed from the line by the resonating molecules,

$$I' = \int_{-x_1}^{-x_1} Ce^{-kx}dx = \frac{2C}{\sqrt{k}}\int_0^{\sqrt{k}x_1} e^{-u^2}du,$$

or for the ratio

$$\frac{I'}{I} = \frac{2}{\sqrt{\pi}}\int_0^{\sqrt{k}x_1} e^{-u^2}du.$$

Now the value of $\frac{I'}{I}$ has been determined by the photometric experiments. For the concentration $\frac{1}{2048}$, or the flame containing the least amount of sodium $\frac{I'}{I} = 0.25$.

From this value and the tables of integrals (Calculus of Probabilities of M. Bachlier) we can calculate the value of the upper limit

$$\sqrt{k}x_1 = 0.0225; \quad \ldots \quad (4)$$

and by division (equations 2 and 4)

$$\frac{x_1}{\epsilon} = 0.27 \quad \ldots \quad (5)$$

We may obtain an approximate value of x_1 if we take for ϵ the value given for a feeble sodium flame by Fabry and Buisson,

$$\epsilon = 0.04 \text{ Å}.$$

Inserting this value in equation (5) gives us

$$x_1 = 0.0108 \text{ Å},$$

or, since x_1 is the half width of the region required, for the width of the region effective in exciting resonance,

$$2x_1 = .021 \text{ Å},$$

the probable width of the resonance lines in contrast to

$$2\epsilon = .08 \text{ Å},$$

the width of the flame lines.

We thus see that by means of sodium vapor at low temperature we can manufacture, so to speak, light much more homogeneous than the incident light, the method being somewhat analogous to that of the Residual Rays of Rubens and Nichols.

It is highly probable that the width of the region removed from the exciting line is identical with the width of the re-emitted resonance radiation. The lines obtained in this way are thus only one-quarter of the width of the lines emitted by the flame and narrower than the iron arc lines.

They are, however, three times as wide as the narrowest known line, the red line of cadmium, for which

$$2\epsilon = .006 \text{ Å}.$$

An interferometer study of the resonance radiation is much to be desired, for the above method of deducing the width of the lines is somewhat circuitous.

No. 14

The Separate Excitation of the Centers of Emission of the D Lines of Sodium

(In collaboration with L. Dunoyer)

The experiments of one of us on the excitation of metallic vapors by monochromatic light have shown that the centers of emission of many spectrum lines are probably in some sort of mechanical or electrical connection. For example, the excitation of mercury vapor by the light of the cadmium spark showed that the vapor emitted the ultra-violet line of wave-length 2536 when stimulated by light of wave-length shorter than any given in the tables at the time, *i.e.*, less than 2000. In the case of the resonance spectra of sodium and iodine we have innumerable examples of associated lines which spring into existence when the vapor excited by light of frequency synchronous with one of them.

In a paper on the 'Resonance Radiation of Sodium Vapor', published by one of us in 1905,[1] it was shown that the vapor of metallic sodium in an exhausted glass tube emitted its characteristic D-line radiation when the image of a sodium flame was thrown upon it by means of a large condensing lens, a cone of yellow light marking the path of the exciting rays through the vapor. If the vapor density was increased the luminosity was restricted to 'a thin skin of yellow light which lined the inner wall of the tube', owing to the failure of the exciting radiations to penetrate the vapor. Precisely similar phenomena were subsequently detected by photography in the case of mercury vapor at room temperature in a bulb of quartz excited by the mercury line 2536.[1]

In the paper on the resonance of sodium vapor, it was suggested that an experiment of great interest would be to excite the vapor by the light of one sodium line only, and examine the reso-

[1] R. W. Wood, *Philosophical Magazine*, November, 1905.

nance-light with a spectroscope: in this way it would be possible to determine whether the two centers of emission could be separately excited.

The experiment appeared, however, to be a difficult one to carry out, and no attempt was made at the time.

Recent improvements by one of us [2] in the method of carrying out the experiment enable a much brighter resonance to be obtained, and make it possible to extend observations over a period of ten or fifteen hours with a single bulb, whereas with the original apparatus the experiment was over in three or four minutes. These improvements have made it possible to carry out at last the suggested experiment on the separate excitation of the centers of emission. The device employed for the removal of D_1 or D_2 from the exciting beam has been recently described by one of us [3] and is an improvement of a method used in an earlier investigation of the anomalous dispersion of sodium vapor.[4]

It is a polarization method, and may be briefly described as follows:—

If plane-polarized monochromatic light is passed through a plate of some doubly refracting crystal with its direction of vibration making an angle of 45° with the axis, it will emerge plane-polarized parallel to the original plane for certain thicknesses of the plate, and plane-polarized at a right angle to this plane for other thicknesses. For intermediate thicknesses it will be elliptically or circularly polarized.

If we employ a plate of quartz 30 mm. thick the emergent waves of D_1 and D_2 of sodium will be plane-polarized at right angles to each other, and either can be quenched by a nicol suitably oriented. If white light is used, and analyzed by a spectroscope, the spectrum will be furrowed by dark bands, the distance between a bright and a dark band being, in the yellow region, 6 Ångström units, the distance between the D lines. As it is necessary to employ a large condenser and work with very divergent and convergent cones of light, a block of quartz of very large size must be used, placed between the two halves

[2] Dunoyer, *Journal de Physique*, January, 1914.
[3] Wood, *Philosophical Magazine*, March, 1914. (This Monograph, No. 12.)
[4] Wood, *Philosophical Magazine*, May, 1912.

of the condenser, since the rays which traverse the block must be parallel. If this is not the case, different pencils will traverse different thicknesses, and will be differently polarized. Moreover, one-half of the light is lost at the start by the polarizing nicol. This difficulty was overcome by employing a large double-image prism, and subsequently analyzing by a double-image prism. In this way, with proper orientation of the prisms, the two images containing only D_2 light were superposed, the D_1 images (of one-half the intensity) lying to the right and left. By this expedient the D_2 image had the full intensity, except for the loss by reflection from the six transparent surfaces of the prisms and quartz block.

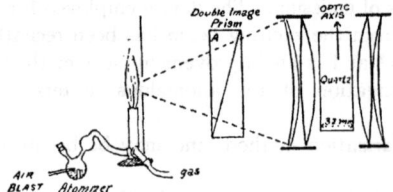

FIGURE 1
(See page 152 for complete figure.)

The quartz block used in the experiment was prepared by M. Bertin from a selected crystal of Madagascar quartz, which was previously examined by sodium light between crossed nicols to make sure that no irregularities of crystallization were present. The block measured 85 mm. × 60 mm. × 32 mm. and gave excellent results.

The arrangement of the apparatus is shown in the diagram.

The double-image prism A is so oriented as to have the vibrations of the two transmitted beams at 45° with the vertical. The second double-image prism is oriented *in the absence of the quartz block*, so as to give two images of the source side by side, and separated by a distance equal to the width of each image. Each image contains both D_1 and D_2. If now the quartz block is placed between the two halves of the Dunoyer condenser, all of the D_2 light (for example) leaves the original images, and unites into a third image between the other two. This is the image utilized. If it is desired to have D_1 light in this image, it is necessary only to rotate the block a degree or two, thereby increasing the optical path in quartz. If we rotate the analyzing double-image prism the central image splits up into two, and the correct position of the prism can be determined by bringing these two images into exact coincidence. This method obviates the 50 per cent. loss of light which results from the use of Nicol's prisms. As the double-image prisms had apertures considerably less than that of the quartz block, they were placed midway between the condenser lenses and the image and source, as shown in the figure. With prisms of too small aperture, however, placed close to the source and image, the separation of the images would be insufficient.

The source of light was a Meker burner operated by an airblast charged with the spray of a very dilute solution of sodium chloride (a saturated solution diluted with 100 parts of water) obtained by means of an atomizer.

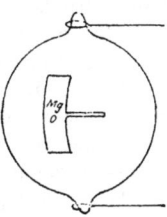

FIGURE 2

The burner was surrounded by an iron chimney with a vertical aperture measuring 25 × 5 mm., these dimensions giving the best results with the polarizing prisms employed. The three images of the aperture, the central one of twice the brilliancy of the two lateral ones, were thrown upon the sodium bulb, which was mounted in the current of hot air rising from a large tube of fire-clay with a Meker burner at the bottom. The sodium bulb was first coated with a deposit of magnesium oxide by holding it above a burning ribbon of the metal. The deposit was then carefully wiped off with the exception of a rectangular patch and a small horizontal strip as shown in Figure 2. This facilitated the adjustment of the polarizing separator and the spectroscope, a constant deviation instrument designed by Broca-Pellin, similar to the type now supplied by Hilger.

The bulb was rotated in its wire supports until the central band of light (the two superposed images) fell upon the strip of magnesia; the collimator of the spectroscope was pointed at the illuminated strip, and a sharp image of the latter formed on the slit by means of a small Dunoyer condenser which is free from aberration if monochromatic light is employed, as is the case in the present work.

The spectrum was now examined with an eyepiece. If both D lines were visible the quartz block was rotated until one or the other completely disappeared. It sometimes happens that only D_2 appears at the top of the image, while both D_1 and D_2 are seen at the bottom: this results from the circumstance that, owing to the finite size of the source of light and the consequent slight obliquity of some of the pencils, the polarizing separator distributes the monochromatic light in bands, resembling interference bands.

If the quartz plate is turned slowly the bands file across the strip of magnesia. The brighter bands represent D_2 and the feebler D_1. If the two sodium lines were of equal intensity the bands would of course be invisible.

If, now, these bands are not parallel to the strip, but cut across it obliquely, it is clear that the illumination may be by D_2 at the top, by D_1 and D_2 at the middle, and by D_1 at the bottom. In this way it would be possible to obtain, with one exposure, three types of excitation.

This method was not used, however, and the quartz block was adjusted by tipping it slightly in the direction of the source or image until the bands were vertical and the illumination constant and of the same type (D_2) all along the slit of the spectroscope. The bulb was now rotated until the strip of magnesia fell into coincidence with one of the lateral images (D_1), the narrow horizontal strip of magnesia cutting across the D_2 image. The burner which heated the bulb was now lighted, and as soon as the superficial resonance reached its full intensity the exposure was commenced. Panchromatic plates (Wratten & Wainwright) were cut into small squares large enough to cover the eyepiece tube of the spectroscope, and were held in place against the latter by two thicknesses of black cloth and a rubber band. This method permits of very accurate focussing and is extremely convenient if the spectroscope has no plate-holder.

We first used a spectroscope furnished with cinematograph lenses of large aperture (F 4), which was loaned to us by M. Debierne. With a Rutherford compound prism this instrument resolved the D lines on the photographic plate if the slit was made exceedingly fine. The first photograph, which was made of the resonance excited by the light of the D_2 line, showed no trace of

D_1. The exposure was of three hours' duration. A second attempt with a five-hour exposure showed a trace of D_1, but examination of the exciting light showed that D_1 was present. This was found to be due to the rise of temperature (5°) of the room during the exposure, the polarizing separator being fairly sensitive to temperature changes. The first pictures were made before the expedient of the narrow horizontal strip of magnesia had been adopted. This proved to be a great convenience, for a record was left of the integrated condition of the exciting line for the whole exposure.

We finally substituted a large constant deviation spectrograph for the smaller instrument, as the latter barely resolved the lines, and it was often difficult to be sure of what we had on the plate.

This instrument showed very clearly that the D_2 center of emission could be set in vibration without disturbing the D_1 center, in other words we can have sodium vapor emitting *one D line only*.

Photographs made of the spectrum of the resonance radiation excited by both sodium lines (with the polarizing prisms and quartz block removed) showed that the D lines had the same intensity, in some cases D_1 even appearing slightly brighter than D_2. It was found that if the amount of sodium in the flame was reduced to the least amount consistent with having resonance radiation of sufficient intensity to photograph, the D_2 line came out stronger than D_1 in the spectrum of the latter, as is always the case with the sodium flame. Exciting the vapor by the light of a bright soda-flame gave a resonance radiation in which D_1 came out stronger than D_2, which is never the case with the flame. This is due to the circumstance that with a bright flame, D_2 is more or less reversed, hence it is less effective in exciting the resonance, for the vapor in the glass bulb responds only to the wave-length forming the core of the line. D_1 is less easily reversed, and may consequently be more efficient in exciting resonance.

On Plate IX, Figure 1, we have a photograph of the single line (D_2) emitted by the resonating vapor, the greater intensity at the top being due to the light from the horizontal narrow strip of magnesia. Immediately below this (Figure 3) we have the two D lines as emitted by the flame. Figure 2 shows the spectrum of the resonating vapor when excited by both sodium lines from a strong flame, and we find D_1, which is to the left, slightly brighter than D_2. In Figure 4, the resonance was excited by a feeble flame and D_2 is brighter than D_1.

This change in the ratio of intensity of the two D lines of the resonance radiation leads to some curious results which were somewhat puzzling at first.

It will be remembered that the central patch of light furnished by the polarizing separator results from the superposition of two images, and, other things being equal, will have double the luminosity of the two lateral patches adjacent to it. If the light of the D_2 line, which is brighter than D_1 in the flame, is thrown into the central patch, we should expect it to be more than twice as bright as the lateral patches adjacent to it. It was observed, however, that the resonance radiation from the central patch was often no brighter than that from the lateral patches which were excited by D_1 light. It was found, however, that if the amount of sodium in the exciting flame was reduced, the lateral spots of resonance radiation diminished in intensity, while the central one changed scarcely at all, retaining a brilliancy of about double that of the lateral ones. This will be easily understood from what has just been said about the greater intensity of the D_1 line in the resonance radiation excited by a brilliant flame.

Photographs showing this phenomenon are reproduced on Plate IX, Figures 5, 6 and 7. Figure 5 is a photograph of the three patches of exciting light thrown on a bulb covered with magnesia. The central one, which contains the D_2 light, is twice as bright as the lateral ones (D_1).

In Figure 7 we have the photograph of the resonance radiation from the bulb under the same conditions of illumination. All three strips have approximately the same intensity. Figure 6 was made under the same conditions, except that the amount of sodium which the atomizer fed to the flame was greatly reduced. Here we have practically the same intensity ratio in the case of the resonance radiation as obtains in the case of the magnesia. All of these results are easily explained by the circumstance that D_2 is more easily reversed than D_1. The white spots of light in Figures 6 and 7 are due to regular reflection from the glass walls of the bulb.

The light of the exciting flame was examined with a very powerful echelon of 20 plates in optical contact, each plate 15 mm. in thickness. This instrument was loaned by the kindness of Mr. Twyman of the Adam Hilger Co. The D lines were examined separately by interposing the polarizing separator between the flame and the instrument. It was found that D_2 showed a faint trace of reversal, even with the minimum quantity of sodium in the flame. D_1, however, reversed only when the amount of sodium was considerably increased.

The resolving power of the echelon was about 300,000, and, judging from the ratio of the width of the lines to the distance between the spectra of adjacent orders, the total observable width was about 0.13 Ång.

The absorption of the vapor in the glass bulbs was also examined with the echelon, employing the flame as a source. A distinct increase in the reversal of the D_2 line was observed, when the temperature of the bulb reached 120°. The diameter of the bulb was only 5 cm., and it is probable that with an absorption-tube one meter in length the absorption could be detected at a temperature not very much above the melting-point of the metal.

No attempt was made to photograph the spectrum of the resonance radiation excited by D_1, as it is quite certain that, if the frequency of D_2 does not give rise to D_1, the same will hold true for D_1, as in all cases of resonance spectra the wave-lengths longer than those of the exciting light are much more intense than the shorter ones.

The mechanisms which produce the D_1 lines are not, however, isolated completely; for it has been shown by one of us [1] that excitation of the vapor in the region of the channelled spectrum by means of blue-green light causes the appearance of the D lines in the emission spectrum, or at least of a yellow band which coincides with the position of the D lines. This band may, however, correspond to a curious band spectrum which is symmetrical about the D lines which appeared in the spectrum excited by the cathode rays.

[1] Wood, *Philosophical Magazine*, x., p. 408 (1905).

No. 15

Resonance Radiation of Sodium Vapor Excited by One of the *D* Lines

(In collaboration with F. L. Mohler)

The results given in the previous paper indicate that the centers of emission which radiate the *D* lines can be separately excited.

Other resonance phenomena show that the radiation centers in sodium are not entirely independent. One of us [1] showed in 1905 that excitation of sodium vapor by blue-green light, in the region of the band spectrum, gives rise to the *D* lines, or, at least, to a band in that region.

Stutt [2] in 1915 found that resonance radiation consisting of the *D* lines could be excited by the 3300 doublet of sodium, the second doublet in the principal series of which the *D* lines are the first. When only one line of the 3300 doublet was excited by a coincident zinc line both the *D* lines again appeared. This remarkable discovery, indicating clearly some connection between the emission centers of the doublets of the principal series of sodium, made a further study of the excitation of resonance by one of the *D* lines seem desirable. In view of results which will be mentioned further along, it may be well to point out that Strutt's results may have been due to the presence of hydrogen in his bulb of sodium vapor.

The arrangement of the apparatus and the method used in the present work differ only in minor details from the method of Wood and Dunoyer.

The chief requirements for the investigation are:

(1) A method of completely separating D_1 and D_2 in the exciting light with the least possible reduction in the intensity of the light.

[1] Wood, *Philosophical Magazine*, [6], 10, 408, 1905.
[2] R. J. Strutt, *Proceedings of the Royal Society*, Series A, 91, p. 511.

(2) The preparation of bulbs containing sodium vapor that will give brilliant resonance during a prolonged heating.

(3) The analysis of the light by a spectroscope giving the greatest possible intensity of light commensurate with the dispersion necessary to clearly resolve the D lines.

THE METHOD OF SEPARATING THE D LINES

The large quartz block measuring $85 \times 60 \times 32$ mms., prepared for the experiment of Wood and Dunoyer, was refigured and supported rigidly in a brass frame arranged to rotate on an axis parallel to the optic axis of the crystal, and inclined at an angle of $45°$ to the vertical. It was placed between two large Iceland-spar prisms of about the same size as the block of quartz and mounted with their edges (optic axes) vertical. The lenses of a large Dunoyer condenser made the light passing through the prisms parallel, and brought it to a focus on the bulb of sodium vapor.

This optical system was enclosed in a wooden box, which was kept at a constant temperature to within $0.1°$ C. by a benzine thermostat. This precaution is necessary for a change in temperature of a degree or two will completely change the nature of the light transmitted by the quartz block. A long handle fastened to the supporting rod of the quartz block made it possible for an observer at the spectroscope to turn the block and adjust the apparatus for the extinction of one of the D lines.

THE SPECTROSCOPE

For analysis of the resonance radiation a large two-prism spectroscope furnished with portrait objectives of three-inch aperture and twenty-four-inch focus, was employed. With this instrument brilliant illumination and clear resolution of the D lines were secured with a fairly wide slit, though the definition was not perfect. This spectroscope had been arranged for use as a monochromator, with the second slit mounted on a screw, so that it could be moved along the spectrum. For the present work the photographic plates were simply clamped against the second slit mounting. This offered a very convenient method of taking a series of exposures on the same plate, side by side instead of one above the other, as in the ordinary plateholder.

LIGHT SOURCE

The source of light was a Meker burner surrounded by a chimney provided with a rectangular aperture measuring about 2×5 cms. The image of this rectangle, formed by the polarizing separator, was thrown on the bulb of sodium vapor. A disk of asbestos soaked in salt solution touched the edge of the flame and this disc was revolved once in twelve hours by the hour hand gear of a clock. This device kept the sodium flame at about the proper intensity to give the maximum brilliancy of resonance.

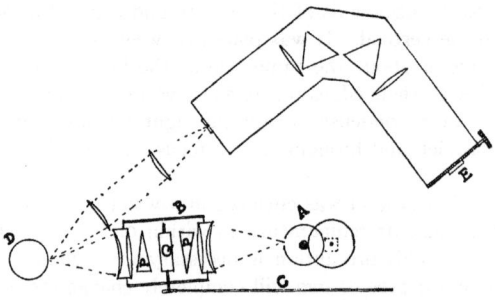

FIGURE 1
Arrangement of Apparatus

It is very important, however, to have the disk graze the flame on the side furthest removed from the lens, as by this arrangement reversal of the D lines is obviated. This is of fundamental importance since the resonance radiation is excited by the core of the line only.

HEATING DEVICE

The bulb containing the sodium was supported above an asbestos chimney about two feet high and five inches in diameter, below which was placed a nest of Bunsen burners. The bulb was supported by a wire frame in such a way that it could be turned about a vertical axis, and a firmly supported pin point touched the front surface to detect any possible displacement when the bulb was rotated.

Fig. 1 shows a plan of the apparatus. The light source, asbestos disc, and surrounding chimney are at A. B is the optical system for separating the D lines with the long handle C to turn the quartz block Q between the spar prisms PP. The bulb of sodium vapor is at D, placed in the position shown so as to prevent directly reflected light from falling on the slit of the spectroscope. At E is the plateholder of the spectroscope mounted on a horizontal screw.

The improvements over the apparatus previously used are in the device for separating the D lines and in the spectroscope. The spectroscope gave better illumination and the system for separating the D lines gave almost perfect extinction of D_1,

FIGURE 2

though the extinction of D_2 was not quite as good, as D_2 has double the intensity of D_1 in the case of the comparatively feeble flame used for the excitation.

PREPARATION OF THE BULBS

The method used in preparing the sodium bulbs is practically that described in the preceding paper. A bulb about 5 cm. in diameter is made as shown in Fig. 2. A piece of sodium, weighing about .2 of a gram, is put in the tube at the left, the tube immediately sealed at A and the bulb connected to the pump and exhausted. The bulb is heated for about half an hour to free the glass from occluded water and the sodium is then distilled into it and the side tube sealed off. The sodium is then distilled from one side of the bulb to the other many times by heating opposite sides alternately with a Bunsen burner, while the pump is kept running and the pressure read from time to time on a McLeod gauge.

The preliminary heating prevents or, at least, retards the reaction of the sodium with the glass which at temperatures above 200° reduces the silicon oxide and makes the glass brown and finally opaque. Bulbs of Pyrex glass, which proved to be far superior to ordinary glass in this respect, prepared in the way described, showed scarcely any color after twelve hours' heating at 220°, and were quite transparent, though brown, after heating twelve hours at 300°.

The repeated distillation of the sodium was to drive off the hydrogen which is occluded by it in large quantities. If, after distilling the sodium into the bulb the pump was cut off and the sodium driven from one side of the bulb to the other two or three times, the pressure gauge indicated a rise of about 0.3 mm., and repeated distillation with the pump maintaining a vacuum of about 0.002 mm., only removed this hydrogen very slowly. The sodium vapor seemed to carry most of the hydrogen with it as it was distilled from side to side of the bulb, for when the pump was stopped and the bulb heated the pressure always increased several hundredths of a millimeter. In the preparation of one bulb the sodium was distilled back and forth across the bulb 170 times in a high vacuum after which the pump was stopped and 0.01 mm. of gas was given off by the sodium when it was heated. In cases in which it was desirable to have some hydrogen present the pump was cut off as soon as the sodium distilled into the bulb. The bulb was then heated, the pressure measured, and the bulb sealed off. To test whether prolonged heating increased the amount of gas some bulbs were opened under mercury after they had been used, but the amount of gas present was not noticeably different.

The careful removal of all the hydrogen is not necessary to secure brilliant resonance, but it does affect the character of the resonance spectrum, as will be explained later.

PROCEDURE

To facilitate the adjustment of the apparatus for extinction of one of the D lines a patch of magnesium oxide was put on the surface of the bulb by burning magnesium wire below it and then removing all the oxide except a small rectangular strip. To adjust the apparatus the bulb is placed in position and turned until the light from the sodium flame falls on the patch of oxide. As magnesium oxide is a nearly perfect reflector this gives a source of light bright enough to make all adjustments. To photograph the resonance spectrum the bulb is turned slightly till the exciting light falls on clean glass. Care must be taken that no light is directly reflected into the spectroscope.

Owing to the path difference through the quartz block of rays coming from different parts of the rectangular aperture, the illumination is not strictly monochromatic (D_2) over the entire image of the aperture. Experiments showed that we have pure D_2 radiation along slightly curved and nearly vertical strips two or three millimeters in width, the distribution of the illumination being somewhat as shown in Figure 3. The upper and lower portions of the image of the patch of resonance radiation thrown on the slit were excited by both lines when the central portion was excited by one only. For this reason any motion of the image on the slit either during the exposure, or in turning the bulb before the exposure, had to be guarded against. As in some cases only a small part of the line was single, it was necessary to compare corresponding parts of the lines of the exciting light and of the resonance light. This comparison was facilitated by the possibility of making several exposures on the same plate with the lines side by side.

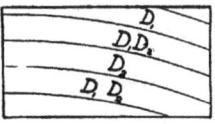

FIGURE 3

The usual procedure was to first photograph the exciting light reflected from the magnesium oxide, then move the plate, turn the bulb and expose to the resonance light, and at the end of the exposure again turn the bulb and move the plate and expose to the exciting light.

The exposures for the resonance spectrum varied from three to fifteen hours; usually twelve hours. The exposures for the diffusely reflected exciting light, to give the same intensity as the resonance light in twelve hours, were from fifteen to thirty minutes when the same type of flame was used. The brightest resonance is secured when the flame is quite faint. Wratten and Wainwright panchromatic plates were used.

The method of estimating the intensity ratio of the D lines, when both appeared, was to match the two lines with sodium lines on a comparison plate made by taking a series of exposures of varying length with a sodium flame of constant intensity. The intensity ratio was assumed equal to the ratio of exposure times of lines that matched.

RESULTS

Most of the plates taken were of resonance excited by D_2, for as D_2 is about twice as bright as D_1, there are obvious advantages in trying it first. The efficiency of the polarization method of cutting out D_1 was tested and it was estimated that under the best conditions D_2 was at least fifty times as bright as D_1, though overlapping due to irradiation of the D_2 line made it impossible to be sure of the ratio.

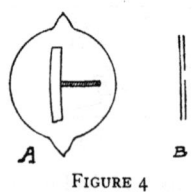

FIGURE 4

The results of many exposures to resonance excited by D_2 showed visible traces of D_1 in nearly every case, but with an intensity ratio of D_2 to D_1 that varied from about six to one, to about twenty to one. This result led at first to the suspicion that stray sodium light was in some way thrown on the spectroscope slit. All possible precautions against this source of error were taken.

When with these precautions both D lines appeared in the resonance spectrum a further precaution was taken to be sure the effect was not false. A narrow horizontal strip of magnesium oxide was placed so as to intercept the rectangle of resonance light in such a way that part of the resulting spectrum line was formed by resonance, and part by reflected light. Since the resonance light is much fainter than the light reflected from a white surface the comparison strip was made a dark gray by first coating the bulb with smoke, and then depositing magnesium oxide until the reflected light was of about the same intensity as that of the resonance. Figure 4 A shows the form of the oxide patch the large rectangular strip being the same as that before mentioned; and the narrow shaded strip the part that intercepted the patch of resonance. Figure 4 B shows the appearance of the resulting spectrum line as it appeared when this method was used. Both D lines appear except at the place where the exciting light is reflected from the gray strip into the spectroscope, and there only one line is recorded. If the appearance of both lines was due to reflection of stray light from the surface of the glass there would be no break in the line. This effect was found both with resonance excited by D_2 and by D_1, and seemed to be conclusive evidence that D_2 light did excite a trace of D_1 light in the resonance radiation. Having verified the results by this method the gray strip was dispensed with in the later work, as it somewhat complicated the adjustment of the bulb. On all plates, however, three exposures were taken, as is shown on Plate X. The plates have been enlarged about ten times. In each case the middle line, or pair of lines, is the resonance spectrum, and the two lines on each side are of the exciting light diffusely reflected from the patch of magnesium oxide, one taken *before* the exposure to resonance radiation, the other after. False effects due to any change in the exciting light can thus be detected.

Comparison of a number of plates taken under different conditions of temperature, with bulbs prepared in different ways, did not at first show clearly under what conditions D_1 appeared in the resonance spectrum excited by D_2. This was due to the fact that two causes contributed to the effect. However, all the bulbs from which the hydrogen was not carefully removed showed D_1 distinctly. Now the resonance spectrum of iodine vapor excited by the green mercury line is changed in the presence of electro-positive gases such as helium and hydrogen, the effect of the gases being to transfer energy from the radiation centers directly excited by the mercury line to other radiation centers. The effect will be described later. The possibility of a similar effect in the case of sodium resonance led to the following experiments: The effect of a change in the density of pure sodium vapor on the resonance excited by D_2 was first investigated. A bulb containing sodium that was as free as possible from hydrogen was used for three exposures to resonance excited by D_2 at temperatures of 210°, 270°, and 340° keeping all other conditions constant. The exposure at 210° showed no trace of D_1 in the resonance spectrum, while at 270° a distinct trace of D_1 was seen. At 340° the intensity ratio of D_2 to D_1 was about the same as that in a faint flame, about two to one, but the plate was much underexposed and the result, therefore, was subject to error. Plate X, Figures 1 and 2 show the resonance of pure sodium vapor excited by D_2 at 210° and 300° respectively.

It may be well to mention here the change in general appearance of the resonance as the temperature is raised. Resonance light becomes visible at about 120°, and appears as a faint glow throughout the bulb. As the temperature is raised the light becomes brighter at the front surface and fades out in the interior of the bulb until, finally, the light is limited to the surface and exhibits a sharp image of the source when it is focused on the bulb. At 200° the resonance appears only at the surface though the image of the source is still a little indistinct, but above 250° the image is as sharp as if the light was reflected from a piece of smooth paper.

The change in the resonance spectrum when hydrogen was put into the bulb was more marked than the change when the vapor pressure of the sodium increased. The resonance excited by D_2 in a bulb containing 0.25 mm. of hydrogen showed D_1 about a quarter as bright as D_2 at 210° and at 300° D_1 was a third as bright as D_2. Figures 3 and 4 were taken under these conditions. The faint line in Figure 3 is of no importance.

A similar series of exposures was taken of the resonance excited by D_1. It is difficult in this case to avoid traces of D_2 in the exciting light for reasons before mentioned, and the intensity of the resonance is reduced to about half. The effect of increasing the vapor pressure or putting hydrogen in the bulb is the same in this case as with D_2 excitation, though the intensity ratio of D_2 to D_1 with D_1 excitation is greater than that of D_1 to D_2 with D_2 excitation when other conditions are the same. Thus with pure sodium at 210° a trace of D_2 was visible (Figure 5) while with 0.1 mm. of hydrogen in the bulb D_2 is half as bright as D_1 (Figure 6). Some plates, where more hydrogen was in the sodium bulb, showed D_2 nearly as bright as D_1 but the plates were under-exposed and there was a possibility that the effect was false.

All the results mentioned above were verified by repetition of the experiments. In all about fifty plates were taken in which the resonance lines were distinct, and the other conditions favorable as far as could be ascertained.

Estimates of the intensity ratio of the D lines were made in each case. They agreed roughly under apparently similar conditions of vacuum, temperature of bulb, etc., but there was quite a range of uncertainty both in the estimation of the intensity ratio, and in the ability to get conditions identical in two cases. The observations may be summarized as follows:

D_2 EXCITATION

Bulb as free from hydrogen as possible.

At 210° (no trace of D_1) intensity ratio of D_2 to D_1 at least twenty to one. Figure 1.

At 300°, ratio of D_2 to D_1 five to one. Figure 2.

Bulb containing about 0.25 mm. of hydrogen.

At 220°, ratio of D_2 to D_1 four to 1. Figure 3.

At 300°, ratio of D_2 to D_1 three to one. Figure 4.

D_1 EXCITATION

Bulb free from hydrogen.

At 220°, a trace of D_2 seen. Figure 9.

At 300°, ratio of D_1 to D_2 three to one.

Bulb containing 0.1 mm. of hydrogen.

At 220°, ratio of D_1 to D_2 two to one. Figure 10.

Bulb containing 0.25 mm. of hydrogen.

At 250°, ratio of D_1 to D_2 possibly three to two.

CONCLUSION

As it has been shown that the presence of hydrogen causes both D lines to appear when resonance is excited by one D line only, it is safe to conclude that the appearance of both D lines at high temperatures is due to the increase of the pressure of the sodium vapor. From the measurements of vapor tension by Hackspill[3] we can estimate the pressure of sodium vapor at the temperatures used. Extrapolating the vapor-tension temperature curve given by him gives the following values of vapor pressure:

> At 200°, 0.003 mm.
> At 250°, 0.01 mm.
> At 300°, 0.025 mm.

Thus at 200° the vacuum is nearly as good as in a cold bulb, but at 300° the amount of sodium vapor is comparable to the amount of hydrogen present, in the bulbs made to show the effect of that gas.

[3] Hackspill, *Annales de Chemie et de Physique*, 28, 680, 1913.

There is a striking analogy to this effect of hydrogen and sodium vapor on the resonance spectrum of sodium, in the case of the resonance spectrum of iodine vapor excited by the green mercury line when traces of a chemically inert gas are present.[4]

We conclude that the transfer of energy from the D_2 to the D_1 emission centers, or *vice versa*, is in some way the result of molecular collision, either of sodium with hydrogen or of sodium with sodium. It has been shown that hydrogen and sodium vapor, both electropositive, cause this transfer of energy, and the analogy to the similar transfer in the case of iodine resonance is of considerable interest.

In a bulb of pure sodium at 220° the surrounding vapor is not dense enough to have an observable effect on the radiation centers, and only one line appears in the resonance spectrum. The appearance of the other line results from an increase in the collision frequency, which increase can be caused either by the introduction of hydrogen at low pressure or by increasing the density of the sodium vapor.

[4] Wood and J. Franck, *Philosophical Magazine* [6], 21, p. 265.

Bei Fragen zur Produktsicherheit wenden Sie sich bitte an:
If you have any questions regarding product safety,
please contact:

Walter de Gruyter GmbH
Genthiner Straße 13
10785 Berlin
productsafety@degruyterbrill.com